To Esther, Joe, Nathan and Aaron
with love for all time.

CONTENTS

PREFACE AND ACKNOWLEDGEMENTS

The subject of time is vast and yet there appears nowhere an explanation of what it is. It seems that only the effects of time can be felt and described, effects containing an element of temporality. It may indeed only be meaningful to discuss time in terms of personal experience. If that is so then this book is a subjective account of my personal exploration of time, even though much of it may appear objective. What I recount is my journey through time, both in the sense of my discovery of elements of temporality in the sciences I know about, physics and astronomy, and also in my personal encounter with the effects of time. So this book combines a mixture of scientific knowledge and human experience.

The book's origins lie in one moment of time, 10.23 p.m., 15 December 1978. That was the moment when my friend John-Francis Phipps asked me to collaborate with him on a book about time, on which he had already embarked. We did indeed collaborate for some months, but as we progressed we both realized that we wanted to write different and separate books, which is what occurred. His is called *Time and the Bomb* and mine is this.

My journey through time since then has taken many twists and turns. Like all journeys, this one has had its false trails, obstacles to be surmounted and thankfully some easier terrain to cross. I have had good guides, both over familiar and unknown country. Some of the places I have been to have been regarded as forbidden territory by many of my scientific colleagues. My path has taken me across the borderlands of science and into the paranormal.

The difficulty for the scientist faced with the paranormal is that he really has to take it all. The consequence of accepting the digestible parts, of stepping onto the foothills so to speak, is that you must also accept the difficult, the inexplicable parts as well, recognise the mountain peaks above you. To do so invariably means altering your whole approach to the world and such a transformation will inevitably be painful and shattering. It requires you to let go of something you think is very precious. With that letting go, however, comes a new creation, a new description of reality which will inevitably be richer than the former one, which seemed so hard to let go of. This has been my experience and that of those whom I know have attempted to climb these particular mountains.

The route I have pursued was not only sometimes difficult but has also missed out many interesting places. The reader will find little in this book, for example, about time in the life sciences or in the social and psychological sciences, despite the fact that both these areas contain much of interest to explore and examine. There is vastly more omitted from this work than will be found between these covers; yet the way has seemed, to me at least, packed with much that is fascinating, some things that are commonplace and some totally strange ideas and manifestations of time. I only hope I have made some sense in linking together these encounters with time.

The material I have collected comes from a wider range of sources than I can easily identify. I have provided a bibliography at the end of the text so that the interested reader can trace from where some ideas come and pursue others in more detail than I have given. I have tried to include in this list all sources I refer to in the text and many more besides that cover particular topics in more depth. Some sources have been used in more than one chapter but only appear in the list in the chapter in which they were first referred.

It is even harder to identify all those people whose conversations, lectures and discussions have clarified my own views and enlarged my own knowledge. Among those to whom I feel a special debt are Angelina Danaid, David Black and the Dominicans at Blackfriars, especially Simon Tugwell and Herbert McCabe, not for specifics so much as for making the context in which I have written so rich. I am indebted to Geoffrey Cornelius both in that last category of people but also for his contribution to my understanding of astrology as well as time. His comments on my manuscript were especially valuable. To him a special thanks. Arthur Oram has also made some useful comments on Chapter Six, which I gratefully acknowledge. I also express thanks to my colleagues and all my students, past and present, for their special and often unknown contributions to my work.

Typing manuscripts (especially knowing how awful my handwriting is) is a thankless task but Carol McCall, helped by Joyce Pickford, coped admirably and helped greatly to get the book completed on time. Thanks to both of you and especially to Carol whose address has provided me with such a lovely example of seriality for Chapter Seven. Also I warmly thank Marie and Dennis O'Malley for putting up with me and putting me up while completing the main part of the writing. Their kindness and critical help was invaluable.

Most of all Farrell Burnett must be thanked for seeing the book through, having faith in it from the start and for editing with patience and the right degree of ruthlessness necessary to cope with the worst side of my writing nature. The book owes enormously to her.

With Time my theme I only have time left to say thank you to my wife and children, whose time I robbed to find time to discover time. This journey was made possible by their generosity and I dedicate it to them, with love.

Michael Shallis
Oxford,
1981.

1

THE ELUSIVENESS OF TIME

Time is ... one damn thing after another; an endless succession of events; time is a measure of change, the separation of events; time is a fleeting illusion. Time is.

The beginning of this exploration of time must be based upon the understanding that time exists. Some philosophers and scientists have questioned even this basic premise. Despite whatever else will be questioned in the course of this examination of an old and difficult problem, to deny time's existence would be to deny human experience.

Time has always fascinated man and yet it is so slightly understood. Time is perhaps the most elusive yet familiar phenomenon we experience. So how can we try to pin it down? This book examines the descriptions of time that emerge from the modern sciences of physics and astronomy as well as from those areas of experience that modern science finds either too awkward or too ephemeral to encompass. This quest is scientific in approach, rather than philosophical, but scientific in its more ancient meaning: a search for knowledge, wherever that search goes, ignoring boundaries or taboos that present custom upholds. Modern science is both expansive and restrictive. Its aim is to encompass everything by its descriptions and analyses but it restricts what it chooses to be 'everything'. In that sense modern science is less about the love and pursuit of knowledge of nature, the natural philosophy of the early modern scientists, who, like Newton, regarded themselves as children filled with curiosity and wonder, but is more concerned with the mastery of nature. The scientific attitude that I try to follow is, I hope, that of the wide-eyed child.

Time is many things to many different people but in all attempts to make some statement about time descriptions are used and that is where the problem of understanding time begins. Descriptions, however necessary for communication, are always limiting, whether they are experiential, scientific or poetic. This book is about some descriptions of time, a most elusive quantity which is very difficult to separate from its description. The language used to describe it contains time-laden words, such as 'event', 'succession', 'etc'. Descriptions and arguments can easily become so circular that the quality of time slips away. So, too, is time bound up in consciousness. Man experiences time, can detect all time's endless changes and yet can he ever be sure it exists of its own right, out there, independent of him? Scientists as well as philosophers are concerned with such questions and indeed the objectivity of time and its separation from consciousness is perhaps one of the central issues in trying to understand time from a scientific viewpoint.

St Augustine summarized the dilemma by saying that he knew what time was until he was asked to describe it; that fits very well with how time is experienced. Contemplation on what St Augustine said would lead to the end of questions about time and yet curiosity always starts prompting those questions again!

Despite time's elusiveness, descriptions and attempts at understanding it can be revealing and rewarding even if they do not succeed in pinning it down. By describing time and asking questions of it in various ways a great deal can be learned about aspects of reality that might not have revealed themselves so clearly. Our understanding of reality, or a description of it, will be altered by pursuing time and I hope the quest in search of time in this book will reveal at least something about the way such questions are posed and how such descriptions are formulated. If time is still elusive in the end, which I fear is inevitable, then at least I hope the journey will have been both enjoyable and instructive.

Time is experienced in two fundamental ways. It seems to flow – the passing seconds, days and years – very much like an endless stream or river, with its own unrelenting inertia carrying everything, including us, along with it. Time is also perceived as a succession of moments with a clear distinction between past, present and future. The past and future are connected by the moment of now, the knife edge of experience between what has happened and what will happen. Of course, the moment of now is also connected with the idea of flow, but in this way of describing time the observer is positioned at a static instant and each moment is marked off as time flows past.

Related to these conceptions of time are the more generalized models of linear and cyclical time. Time seems linear most of the time. It stretches back into the past like a temporal ruler marked in a scale of years, decades and centuries, and it stretches away into the future. This view of time arises essentially from the Judeo-Christian tradition, which is typified by the creation, when time began, and distinguished by specific and unique events, such as the birth and death of Christ, and whose end will come in the future. Such a view of time is mirrored in modern cosmology which places the origin of the universe at a specific event, the so-called Big Bang. However, time is also perceived as cyclic and therefore not necessarily 'progressive'. In such a view, based on the various cyclic characteristics of nature, the day, the season, the year, time becomes the element within which natural events occur, always coming full cycle and returning to their origin as spring gives way to summer, autumn and then winter; the eternal, golden braid, as Hofstadter has called this interweaving of complex, cyclic patterns.

These two attitudes to time are analytic (the linear model) and organic (cyclic time). Linear time is dissected into its instants or moments, while cyclic time grows. Marshall McLuhan connects the linear conception of time with language; the word and the sentence are linear in form, analytical, consequential, progressive. In pre-literate society, where time was viewed cyclically, language too was more 'organic' and quite unlike modern western languages. Language cannot be avoided when something is to be described, but the limitations of language can be made explicit. Thus, a 'natural' view of time will be closely connected with its system of description.

In past ages man's perception of time consisted of a sense of the interplay of constantly recurring events; the rising of the sun, the yoking of the oxen, the flight of birds, the tides, the seasons. Time intervals were reckoned in the pacing of activities; the time it took to walk to the next town, how long water took to boil and so on. For example, in India the shortest interval of time that could be described was the time it took rice to boil, about thirteen minutes. Both personal and social life was governed by the hours of daylight and darkness and the seasons of the year and religious festivals. Holidays were taken by the institutions to coincide with harvesting in the late summer, meals were eaten when people were hungry. This interweaving of temporal patterns was described beautifully by Joseph Weizenbaum:

Cosmological time, as well as the time perceived in daily life, was therefore a sort of complex beating, a repeating and echoing of events. Perhaps we can vaguely understand it by contemplating, say, the great fugues of Bach. But ... we must not think in the modern manner of Bach as a 'problem solver', ... instead we must think that Bach had the whole plan in his mind all the time, that he thought of the 'Art of the Fugue' as a unified work with no beginning and no end, itself eternal like the cosmos. . . . We might then find it possible to think of life as having been not merely punctuated but entirely suffused by this kind of music, both on the grand cosmological-theological scale and on the small day-to-day level.

The introduction of the mechanical clock changed all that, and enabled man to perceive time in terms of hours, minutes and seconds. Lewis Mumford wrote that the clock 'disassociated time from human events and helped create the belief in an independent world of mathematically measurable sequences: the special world of science.' The clock modelled an abstract concept of the motions of the sun, moon and earth, and replaced the organic rhythms of nature by the mechanical ticking of its clockwork, the chiming of its bells. With the introduction of the clock to life in the middle ages came a changed perception of time. As Weizenbaum wrote, 'the clock created literally a new reality . . . that was and remains an impoverished version of the old one.'

Now the oxen were yoked when the clock struck seven, now people ate when the clock struck one and not when they were hungry. The new reality, the new experience of time was quantified, mechanized and removed from nature. Temporal experience was impoverished by the clock and the experience of time today is even more impoverished, even more mechanized, even more dependent on the state of the clock rather than on the direct experience of the quality time has. Of course, clocks of one sort or another had been in use long before the mechanical clock was introduced to medieval society, but the regulation of life around the clock and its sounding bell was as great an upheaval as was the invention of printing.

The clock yoked mankind to a linear, countable time and lost for the people of modern Western culture that organic feel for time's endless patterns. The experience of time is circumscribed by the age in which we live, of our time, as it were. Nevertheless, some vestige of that earlier awareness of cyclic time can still be sensed, people still do react to the cycle of the day, the week, the year. It is just that an effort has to be made for such experiences to become conscious, they no longer come 'naturally'.

The existence of clocks brings an awareness of time in a special sort of way. The mind processes the information from clocks and 'interprets' that information as 'being time'. When I first bought a digital watch I was very aware, especially for the first few days, how different time felt, because the sort of processing required of digital information is different from the processing done on analogue information. My mind had to operate differently in connection with the new watch. The language had changed so time felt different in quality.

The problem of time has always existed for scientists. Among the ancient Greeks Zeno was posing his paradoxes of time and motion, for time arises most immediately and explicitly in the idea of motion. If something moves it is changing its position in space in a certain time interval. Motion cannot be described without time. One of Zeno's paradoxes involved describing the motion of an arrow. In order to describe its speed through the air the distance it travels must be divided by the time it takes. To find its instantaneous speed we must reduce the time interval to zero, in which case the arrow does not move at all, its motion is comprised of motionless moments. Thus movement does not take place! And yet things move, things change their positions in a multitude of ways: if change is recognised so must be time. If nothing ever changed would time cease to exist? If so, then time would be purely a characteristic of change. Such a relationist view of time was expressed by the great mathematician Leibnitz in the late seventeenth century.

I hold space to be merely relative, as time is. I hold it to be an order of co-existences, as time is an order of successions. Instants, considered without the things, are nothing at all, they consist only in the successive order of things.

Time as the separation of events, their order and a measure of the duration between them is clearly connected with change, for by 'things' Leibnitz means distinguishable states and for two states to be clearly distinguishable a change must have occurred from one to the other. The relativistic view of time is very much tied up with the changing conditions of things. Such an attitude, of course, implies that time is not an independent, objective characteristic of the world, but merely a way of describing relations between events. It is a view clearly opposed to the absolutist attitude of Isaac Newton, for whom time was very definitely a real thing.

Absolute, true and mathematical Time, of itself, and from its own nature, flows equably without relation to anything external, and by another name is called duration.... All motions may be accelerated or retarded, but the flowing of absolute time is not liable to change.

Newtonian time, like Newtonian space, was absolute. All events could be considered to have a distinct and definite position in space and to occur at a particular moment in time. This approach feels intuitively correct. It matches human experience very well and it seemes unfortunate that attempts at measuring such an absolute system result in problems that force a move from the absolutist to the relationist camp – but that will be dealt with in the following two chapters.

Modern physics is always said to have started with Galileo and Newton and the reason for this is that it was they who attempted to tackle the problem of describing motion for the first time since the Greeks. Galileo's reputed demonstration that all masses fall to the ground at the same rate was the empirical test of his ideas on motion, but he did not achieve more because of his lack of suitable mathematics with which to handle his ideas. What Newton did was to extend Galileo's ideas, combine them with Kepler's studies of planetary motion, and re-examine the problem of motion. His achievement was considerable, and he devised new mathematical techniques to describe change. The infinitesimal calculus, which incidentally Leibnitz also developed independently at the same time, was Newton's descriptive system for coping with change. His technique was not unlike Zeno's, for he realized that instantaneous speed must be calculated by dividing distance by time, but what he did was to make the time interval very small but never quite zero. In so doing he found that the instantaneous speed of an object is calculated from the slope or gradient of the curve describing the motion. Whenever a quantity changes with time a graph can be plotted which will in general be a curve relating the quantity changing to the time taken for each change (see Figure 1.1). The 'slope' of the curve at any point gives the rate of change at that point and Newton's calculus enables the slope or gradient at any point to be calculated exactly. This mathematical technique, then, has built into it the idea of change and hence of time, but rather than deal with time explicitly as some quantity 'out there', it incorporates time in the description of changing events.

In this sense time, or at least change, has been stabilized, made static in its description; the problem of continually moving time has been removed from a description of motion and change. This stablizing power of the calculus made the difficulties associated with the description of motion reduce to mere technicality, the handling of mathematics, freeing scientific description from the more general motaphysical problems that up to then had prevented further progress. Newton's calculus, it seems to me, is relationist in spirit, in that it is concerned with the order and change of events and does not describe the 'absolute' time that he himself envisaged. This is

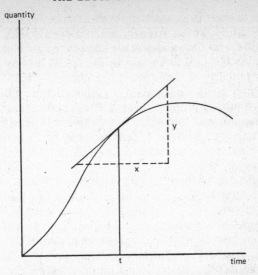

Figure 1.1 *Diagram showing the general curved relationship of a quantity changing with time. The 'rate of change' at any instant, t, is found by measuring the gradient, y/x, of the tangent to the curve at that instant.*

perhaps not surprising, as will be seen later, because absolute time is fine as a concept but difficult to realize in practice.

The invention of calculus provided the basic mathematical tool for the more rapid development of modern science and time was thereby embedded in the matrix of scientific thought in the static form of 'rate of change'. It is interesting to view the historical development of science in the light of the removal of the problem of time. By turning changing time into something unchanging and therefore maneageable, modern science was able to take off after centuries of lack of progress. In by-passing a fundamental problem many practical problems could be solved and this characteristic of science typifies its progress.

Rate of change is a quantity that applies to 'now' and time as such does not flow in the maths or in the overall description of changing events. Although this seems surprising it is not, in practice, a problem because the equations of physics explain the general properties of things and although things change with time, time flow is not a property of these changes but is a property of time itself. In point of fact, time is barely described by physics and it is often left to philosophers and other enquirers to pursue its nature. Indeed, one of the problems with time in science is the lack of a coherent programme to study its properties, for the idea and existence of time is fundamental. However, science, as Medawar has said, is 'the art of the soluble' and time's elusiveness makes the problem of time, perhaps, insoluble, and there may lie the answer to this dilemma.

Another property of time that is brought into scientific thought and discussion is its directionality. Although time's flow is not explicitly handled in physics, there is a recognition that it flows in one direction. Indeed, our ideas of causality and simultaneity are very much bound up with our understanding of time's unidirectional flow. However, in the equations of physics the direction of time, time's arrow, is missing. The equations work as well with time flowing backward as they do with it flowing forward. Of course, scientists understand that time does only flow in one direction, and time reversal would be as absurd in physics as it would be to common sense, so in physics we must look at other properties of the world to find how or why time only flows forward. This aspect of time will be discussed in Chapters Four and Five.

Despite an incorporation of time into a scientific description of the world scientists are still not entirely sure that time exists as such. The separation of time from our consciousness of it is a real problem and one that was faced by a group of eminent physicists, cosmologists and philosophers at a meeting at Cornell University in 1963 to discuss the nature of time. In the opening address Professor Tommy Gold asked:

Is it really basic to the description of physical processes that time progresses? Or is it only that biological mechanisms have devised a representation in terms of flowing time? If this is so then perhaps it is wrong to carry over the concept of time in a description of nature.

After several days of earnest discussion the matter was not resolved. The final summary was given by Professor John Wheeler. The doubts raised at the start of the meeting were still unresolved:

There is a feeling of not really understanding anything. The problem is the 'I' in the question and the sharpness of 'now' that goes with 'I'. Why is it that we have this individuality about experience? The very vividness of 'now' has perhaps something of a psychological character that maybe we had better avoid.

Again, the central problem seems to be one of description and the incapacity to disentangle the world out there from its description. The problem is the 'I' that makes objectivity so very difficult, and to postulate an 'objective' reality, as science must as one of its basic tenets of belief, is an act of faith that can only be justified in certain limited applications. Whether one can postulate an objective, physical time 'out there' is one of the topics of this book. Certainly time is a complex and multifaceted quantity and it may turn out that several of its different faces appear exclusive of one another. This may sound surprising, but although we are normally used to more straightforward descriptions and simple logic, and are most familiar with something either being this or that, contradictory descriptions are not totally unfamiliar.

One well known physical quantity that certainly displays this multifaceted quality is light. When scientists examined the nature of light they discovered it possessed wave-like qualities, in that light displays the phenomena of diffraction and interference that all waves exhibit. However, some properties of light are quite inexplicable in terms of a wave model and can only make sense if light is described as consisting of particles. However, particles and waves are so distinct and dissimilar one ends up asking which one is light, waves or particles? The answer, of course, is both. If we ask a wave-like question of light by performing a wave-type experiment then the result of that experiment demonstrates the wave aspect of light. If we ask a particle-type question of light by performing a suitable experiment we get a particle-like answer (indeed particles of light are called photons). We can never perform an experiment which demonstrates both aspects simultaneously and we conclude that light is such that its nature consists of at least wave-like and particle-like aspects. This wave/particle duality is well understood by physicists and it should not be surprising if other fundamental properties of nature, such as time, also have complex natures. The aspects of those natures that we see and understand are those we ask about, and it may be that in choosing to ask only one type of question limits understanding about the complexity of the phenomenon.

The difference between time and light is that light is described by mathematical equations and time is not. Time enters the equations of physics as a simple entity and its nature is either inherently assumed or is infered from observational evidence. Time is closely linked to motion in the development of physics, indeed one of the first things in physics to be taught are Newton's Laws of Motion that arose from his consideration of time and change. With the idea of motion came the concept of space as well as time. The concept of space is so closely allied to that of time it is impossible to separate them from one another. Objects or events in space are time-laden ideas: 'now' is inherently connected with 'here'. Despite this intimate connection Newton did imagine time to be independent of space. Both space and time, for him, were absolute and all places in his universe could be connected by the same moment of 'now'. But Leibnitz' relationist attitude kept space and time as concepts that merely separated objects and events. There is so much more feel, concretely, for space than for time that it would be easier to deal with time as if it were a form of space. So, if time is taken to be linear and equally divided, it becomes rather like space. Indeed, mathematically, time can be converted to space merely by multiplying it by a velocity. Taking the constant speed of light to be the maximum possible velocity in the universe means that multiplying time by the speed of light turns time into space!

The spatialization of time is a very effective way of by-passing the problem of time in physics and makes the dimensions of time compatible mathematically with the dimensions of space. So, although it may appear to be an odd sort of thing to do, it has practical justification. When Einstein developed the idea of relativity he demonstrated the inseparableness of space and time, adding the dimension of time to the three dimensions of space to give four-dimensional space/time. It is impossible to picture four dimensions, so the convention is often adopted of reducing the three spatial dimensions to only one and adding the time dimension to it to give a two-dimensional spatial picture.

Such a picture is called a space/time diagram and Figure 1.2 is an example. On this diagram an event is pictured progressing from its start at time t_1 and position A and travelling to position B by the time t_2. Such a line on a space/time diagram is called an event line or world line. Figure 1.3 is another space-time diagram showing what Newton's absolute time system is like. On such a diagram it is possible to say that the two events A and B occur simultaneously although they are separated in space. The grid on the diagram consists of vertical lines each of which represents one place at different times, and horizontal lines consisting of different places at the same time.

The use of space/time diagrams is, like mathematics, another way of handling the concept of time for the purpose of description. Like other descriptive systems it has severe limitations and inbuilt assumptions. Clearly some of the assumptions in such a diagram are that time can be spatialized, that it is linear, continuous and connected; that is, it forms a coherent surface. Well, these assumptions are necessary for the mathematical handling of time,

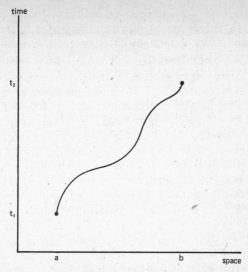

Figure 1.2   *A space/time diagram showing a typical world line for an event moving from a to b over the time interval from t_1 to t_2.*

Figure 1.3   *A space/time diagram illustrating Newtonian absolute space and time. Each vertical line connects one place to all times and horizontal lines connect all places at one particular time. Hence the two instantaneous events A and B occur at the same time although separated in space.*

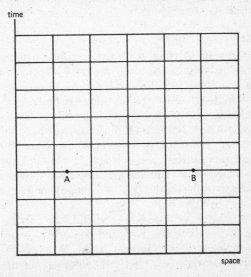

but they do not necessarily resemble the actual qualities of time. Time may certainly display properties of continuity and connectedness. People do not suddenly find themselves at 'another time', neither do they jump from Tuesday to Thursday, missing out the day between. But those properties ascribed to time are *ascribed*, not discovered or revealed. They are part of the descriptive system.

Although time may be spatialized for scientific convenience there is one property of time that is nevertheless different from space and that is its unidirectionality. In space objects can move forward and backward, left and right, up and down, but in time, even spatialized time, they can only go forward. This is a constraint imposed on time and thus in space/time diagrams lines always move upward, forward in time. Also, objects can stop still in space, not moving, but they always keep moving in time: on a space/time diagram an instantaneous event can be pictured as a dot on the illustration, but a line must be drawn if real duration is involved for time never stands still. In later chapters of this book, however, I shall be discussing incidents, experiences and phenomena that run contrary to this picture of time. Experience of timelessness, of eternity, of time reversal, of precognition, of mixed-up times and even of systems operating outside time are all things to be discussed and analysed. Are such phenomena real, are they connected primarily with the psychological perception of time or do they reveal genuine aspects of time that appear contrary to common sense or scientific description? Such experiences should not be dismissed just because they go against common sense. Indeed, science itself reveals some extraordinary aspects of time. Certain phenomena pose paradoxical questions, but wave/particle duality also appears paradoxical, but only because its descriptions run contrary to one another, not because there is a real paradox in nature. Maybe time is embedded in a timeless matrix that we can occasionally perceive. Maybe time runs backward as well as forward but normally only one way of the flow can be detected. Perhaps all times co-exist but the response is inhibited to all other times but 'now' by some biological or psychological mechanism. The further time is explored the more mysterious it becomes – yet the passage from Ecclesiastes which begins, 'To every thing there is a season and a time to every purpose under the heaven' *feels* right.

Such an expression of time is meaningful, and expresses wisdom, and I hope to find both meaning and some wisdom, and not just mundane knowledge, in this search for time. Time represents something of a mystical or archetypal character, separate from the description of time as change in the physical universe. Time has a quality about it, a quality of meaning that is instinctively grasped. To every thing there *is* a season. This quality of time is something apparently outside the physical sciences, and everything that science hopes to understand. The quality of time may be a collective quality of man's expression of himself. It could be that the quality of time influences mankind's collective moods. It is not insignificant that similar discoveries, thoughts, ideas, expressions arise in different people, in different disciplines at the same time. It is not that there are direct influences between people, but that 'something is in the air'. The time has a quality to it. This concept has been formalized, for example, by Rupert Sheldrake in his controversial yet exciting suggestion that this 'something' can be described by a morpho-genetic field, an acausal principle that governs form. Once nature has developed one form it recurs in several places and in different contexts at the same time.

Can a scientific study of time be reconciled with such fanciful discussions? I believe so, and this book is an attempt to make such a reconciliation, to bridge that apparent gulf. I begin with science and the remark that philosophers and scientists vary from each other not in their theorizing but in the fact that scientists eventually have to move outside their theory, and test their ideas in practice, in the physical, rather than the intellectual, rational or logical world. Scientists end up making measurements; this exploration of time begins with the measurement of it.

2
THE MEASUREMENT OF TIME

Time can be defined as that quantity measured by a clock. Such a definition is framed not in terms of the quantity to be measured but in terms of the operation of a measurement technique. For example, length can be defined as that quantity measured by a ruler. Such definitions, much used in science because of their precision and practicality, are called operational definitions.

However, there is a rigidity, a fixedness, about 'time is that quantity measured by a clock'. What sort of clock? Do all clocks measure time in the same way? Do they all tell the same time? What about those aspects of time, like a sense of timelessness, that are immeasurable? This chapter will deal with the question of clocks, but as for the other questions many scientists would answer that any aspect of time not covered by an operational definition is not a scientific question. They might even go on to say that such questions cannot have any meaning. Such purity of thought in science can be misleading to the layperson because many experiences are meaningful but are not and cannot be covered by an operational definition. What this really means is that although such experiences cannot be analysed by scientific methodology they are not invalid as experiences, even though some scientists will argue for their meaninglessness. These two approaches reflect two positions that can be taken about science and scientific thought in general. Some people believe that science is the only form through which we can rationally understand the world, and that nothing else is really real. Others regard science as one of the many ways in which the world can be discovered and understood, but that it is not everything, and indeed, rationality is not everything either, but one of several modes in which human thought can operate. I take this latter standpoint myself, but, as a scientist, that does not prevent me from taking the restrictive but practically necessary operational definition and applying it to its limit.

A word about operational definitions in general. An operational definition such as 'intelligence is that quantity measured by an IQ test' is of course worrying because most people intuitively know what they mean by intelligence and can think of several cases in their experience of people whom they regard as intelligent but who may not have particularly high IQs or, contrastingly, people with high IQs whom are not regarded as being very intelligent. I've used this example quite deliberately to illustrate the need to be wary of operational definitions for two reasons. Firstly, an operational definition need not be true. That is, intelligence may *not* be that quantity measured by an IQ test, although an IQ test may well measure something to do with intelligence. It presumably predicts, for example, people who are likely to do well in examinations, and so on, but it does not necessarily measure intelligence. Secondly, it assumes that intelligence is a single, valued, linear quantity that can be expressed on some suitable scale and implies that there is a norm about which intelligence is more or less evenly spread. These assumptions may not be true, but they allow for a measurement to be made that yields a result that can be used. So, although the operational definition may not match up to what intelligence is, it at least allows a practical measurement to be made which philosophizing does not. The parallels between intelligence and time are many and the comments I have made concerning operational definitions apply as well to that of time. In considering time to be that quantity measured by a clock and in applying that definition to it, this investigation of time may actually not reveal anything about time, but will indicate instead a great deal about this particular view of it, that is, about this description of it.

Since time is perceived as an endless flowing stream, it is not surprising to find that devices in which flow occurs are used as clocks. The hour glass is an excellent example which models several properties of time. As well as modelling the flow of time, it also incorporates the *one way* flow that must be an essential feature of a clock. Time is not experienced in reverse for the same reason that the grains of sand do not jump back to the upper part of the hour glass. The interval of time it takes for all the sand to flow to the lower bulb of the glass provides a convenient unit of duration. The point in time, the moment of now, is noted by observing the state of the hour glass at an instant. The water clock and the burning candle are similar 'flow' devices. The only problem with these clocks is that they do not lend themselves easily to precise quantitative readings. Because their flow is neither regular nor uniform, they are not easy devices to calibrate. Their performance may well be subject to conditions prevailing at the time and place at which they are used, and so, despite their excellent qualitative modelling power, such clocks are not entirely suitable for setting up an operational definition of time.

Time as an endless flowing stream is a linear concept but modelling this flow is not suitable for a practical clock. However, a cyclic view of time, where the repeatability of an event, like the observed passage of the sun through the sky, day after day, can provide a most appropriate basis for a clock. Obviously the day, the month, and the year were the time-keeping devices throughout much of mankind's history, and such natural rhythms are still used to a greater extent than mechanical clocks. But the day is net a clock; and time is not sensibly defined as that quantity indicated by the rising and setting of the sun. The swing of a pendulum, however, makes for an excellent clock. The motion is exactly periodic, adjustable, and can drive a pointer about a dial so that a cyclic property can be converted through a simple mechanism into an indicated flow. The periodic motion of the pendulum sets up a unit of duration and the pointer indicates the moment of now.

Such a clock incorporates cyclic and linear concepts of time, intervals and points in time, one-directional flow and a practical means of quantifying time with reasonable precision. In practice, any regularly oscillating or vibrating mechanism will serve as a 'pendulum', and the balance wheel of a watch driven by a spring, or by an electric battery is but one example. The quartz crystal in more accurate clocks and watches vibrates mechanically, like a pendulum, and sets up a useful oscillating electric voltage for controlling the pointer reading. Even the atomic clock, where the oscillations come from vibrating electrons inside the caesium atom, is a form of pendulum clock: one that reaches an accuracy of about one second in 10,000 years!

Of course, the time interval defined by one oscillation of the electrons in the caesium atoms of an atomic clock is extremely short (about one ten thousand millionth of a second) and cannot in itself be 'timed' by another clock. The caesium clock is regarded as the most accurate and is the standard by which intervals of time are now defined. Its radiation is an indicator of the duration of its oscillation; the wavelength of its emitted light can be measured very precisely, and because that wavelength is constant, the frequency of the oscillations is constant to the accuracy quoted above.

Nowadays the second, the basic unit of time interval, is defined in terms of the number of oscillations given by a particular spectral emission line from atoms of the element caesium (9,192,631,770 in fact), and thus the day and year are defined in terms of 'atomic seconds'. The difficulties in keeping clocks adjusted for 'real' days and years are complicated by the unevenness of the earth's rotation and of its passage about the sun, so that corrections are constantly being made to keep astronomical time and atomic time in step. All the corrections are made in order to keep the time keeping system neat and tidy – time would rapidly become a practical nuisance if clocks were continually drifting out of step.

Of course, historically, clocks were improved to enable navigation and map reading to be refined. In order to tell the longitude of its position a ship would need a clock that was calibrated at Greenwich when it departed on its voyage, so that the timing of star positions could reveal how far east or west the ship had progressed. The accuracy of a clock on board ship was vital. If it lost a few seconds a day then after a week or two the ship could estimate its position quite incorrectly. The sailors would still know when the local noon was, by the passage overhead of the sun, but how that local noon related to noon at Greenwich was the real and practical problem. This difficulty of maintaining a calibration and of having to move clocks around is the key to understanding the difference between a relationist and absolutist view of time, and will demonstrate why the use of the operational definition of time leads to such curious conclusions. But first, standards and astronomical measurements must be examined.

The idea of the day and of the year, the phases of the moon, the cycles through the heaven of the fixed stars and the planets is basic to the experience and understanding of time. Yet nowadays the unit of time is based not on astronomical phenomena but on an atomic property. Of course, as many other writers have explained, the irregularity on a small scale of astronomical cycles and the constant need for minor corrections makes those cyclic properties difficult to use. While the adoption of an atomic clock as the standard eases some of these problems, it also creates another because the atomic clock potentially reveals a different aspect of time from that given astronomically.

With the exception of the atomic clock, other time keeping devices all share the same fundamental property: they operate due to the force of gravity. A pendulum swings because of the gravitational pull of the earth on the bob. The day, the month, the year all demonstrate their cycles because the earth and the moon move around the sun under the pull of gravity. Watches driven by a spring, or by an electrically powered oscillating quartz crystal, are in essence extensions of the gravitational clock: all that has happened is that a spring, or a battery, has been substituted as a local 'gravity' inside the watch or clock, and the time indicated is still calibrated gravitationally.

On the other hand, the atomic clock is based fundamentally on the electromagnetic forces operating inside the atom, and is essentially independent of gravity (although, as will be shown in Chapter Five, all clocks, including atomic clocks, run slow in very strong gravitational fields). If gravitational clocks are based on one physical principle and atomic clocks on another, the question arises: do they tell the same time?

This question is probing something more fundamental than the need to adjust astronomical time to fit atomic clock standards. The scale of the force of gravity is fixed by a universal constant, known as the gravitational constant, which scientists believe has the same value at all times and in all places. (Such faith in the uniformity of phenomena is an essential prerequisite for a sensible science.) But the electromagnetic force is scaled by a different constant. Both these constants are markers that tell us what the size of these forces are under standard conditions. Now if these constants are really constant, that is, if they don't change their value as time passes, then there is no reason why time as indicated on an atomic clock should not be identical with time indicated on a gravitational clock. However, there is some evidence now being examined and tested by scientists that the gravitational constant may be changing with time. The amount by which it changes is extremely small but if it is changing and the electromagnetic constant is not, then the two sorts of clock do not tell the same time; the fundamental processes on which they are based work at different rates, and if duration is measured on each kind of clock, different results will be given. This means, for example, that the age of the universe would be different, depending on the type of clock which determines it. To those who insist that there must be a 'true' age of the universe, this seems nonsensical. On the other hand, how can that age be measured without a clock? So, an operational definition of time must include what type of clock it is that performs the operation. This is yet another limitation on a description of time, a further restriction, making scientific understanding more theoretical, more idealized and less close to experience. And yet it is still very useful and leads to some remarkable conclusions which start to raise questions about the 'objective' nature of time itself.

In order to set up a universal time system, a standard clock must be chosen against which all other clocks must be calibrated. In principle, all such synchronized clocks could be moved to different parts of the universe to form an absolute time system. Everywhere will be the same time at the same time. Unfortunately arranging for clocks to be in synchronization is not that easy.

One of the factors most closely associated with time is motion. Motion means travelling at a speed, and speed is 'distance travelled in some time interval'. Mathematically, speed is expressed as:

$$\text{speed} = \frac{\text{distance travelled}}{\text{time taken}}$$

In order to measure speed, time as well as distance must be measured. That is fine, because a standard clock can be used. But three distinct measurements are necessary. First the distance over which the moving object is travelling, say between two markers A and B, must be measured; secondly, the time at the start of the motion (at A); and finally, the time at the end of the motion (at B) recorded. The difficulty arises because the second time measurement is taken at a different place from the first one. In order to be able to make that measurement, either the standard clock must move or a synchronized clock must be provided at the end position of the motion. Now it may be that moving clocks around could disturb their time keeping qualities (for example, by moving them to a different environment where the force of gravity was different). As this would be disastrous in setting up an absolute time system of synchronized clocks, it might be wiser not to move any clocks around but to synchronize the clock at B by exchanging signals with the standard clock at A. The clock at A could be observed through a telescope from B where that clock could be set accordingly. The signal exchanged between clocks is a beam of light. But light does not travel instantaneously. It has a finite speed. Although the speed of light is very fast (300,000,000 metres every second) it is a definite, fixed and constant speed. How is speed measured? By setting up two clocks and measuring the distance between them! The argument has become circular. In order to measure speed either the clocks must be moved about or synchronized by an exchange of signals. The former method will alter the rate at which the clocks work, and the latter requires a measurement of speed. Of course, a light signal could be sent to a mirror at a known distance, and its round-trip journey could be timed with the same clock. That would overcome circularity of argument and provide a basis for synchronized clocks. The flaw in this argument, however, lies in the fact that it will only give an *average* speed for light. Light, although assumed to travel at a constant speed, might go slowly on the outward journey and go faster on returning, for example. Furthermore, the speed of light is known only to a certain accuracy, so an error or uncertainty will always be associated with a time synchronized by exchanging light signals. The more distant the point B is from A the larger this uncertainty will be. There will still be no way of knowing for certain what the time is at the mirror, unless a synchronized clock is placed there, and that has just been demonstrated as impossible.

This whole argument reveals two important points. Firstly, knowledge of the world, as revealed by science, is limited to what can be observed locally, here and now, and to extend that knowledge requires both a faith in the uniformity of nature and a compromise with truth, for knowledge has an inbuilt uncertainty to it. Secondly, as regards time, perfectly synchronized clocks cannot be set up and therefore there cannot be an absolute time system in the universe, as Newton envisaged, simply because it cannot be measured. Since time is that quantity measured by a clock, and it is impossible to set up a practical system of synchronized clocks, there can be no absolute time. The idea lies outside what can be operationally demonstrated and therefore has no meaning in the purist sense.

But, just because there is no way of measuring it, does that mean that absolute time doesn't exist? The pragmatic answer is that if absolute time can never be demonstrated, it is hardly very helpful to think of time in that sense; alternatively, such a question can be thought of as purely metaphysical and can only be answered in the light of philosophical or religious discussion. However, if absolute time cannot be contemplated, can two events be described as occurring simultaneously? Locally, of course, they can. A statement like 'the postman pushed a letter through my letter box at the same moment I was putting on my coat in the hallway', is a perfectly valid expression of two simultaneous events, but when the two events are well separated in space it becomes much less easy to answer the question affirmatively. For instance, to find out whether an event at some distant location (maybe on another planet) occurred simultaneously with a local event synchronized clocks both here and at that other place would be needed, but it is impossible to make the measurement to ensure their perfect synchronization. The idea of simultaneity assumes absolute time once removed from a very local reference frame, and since absolute time is impossible to determine, simultaneity becomes restricted to local events only. Operationally, simultaneity can be defined as happening when two events occur at the same moment in the history of a clock. Once a second, synchronized clock, or the motion of a clock, is involved, then the definition falls apart. The limitation comes when motion is introduced, and the motion that ends up being the ultimate limit is the motion of light.

Light* is a prerequisite for communication, for the transmission of signals, and its finite speed places several restrictions on what can be done and measured. The speed of light is understood to be constant and all experiments that have ever been made to test its constancy have found it to be so to the limits of detectability. Light travels at an enormous speed which is known with some precision. Methods for measuring its speed have become more and more refined since Römer first produced a 'reasonable' value for it in 1676 by measuring how long it took light to travel across the diameter of the earth's orbit. He observed the four main moons of Jupiter as they moved periodically around that planet, and he deduced correctly that the predicted time for any one moon to reappear from behind the planet would be out of step with the measured time over a six month interval as the light signal had an increasingly long journey to make from Jupiter to earth as our own planet revolved around the sun. In fact, Römer's value for the speed of light was close to the modern value, but more refined methods have been established. Of course, nowadays the speed of light can be measured with pulses of light from lasers with complex electronic detectors coupled to atomic clocks measuring the required time interval, but an important and remarkably precise method was developed by A.A. Michelson in the U.S. in the 1920s. He bounced a light beam off a rotating mirror into a long light path between stationary mirrors, such that when it returned it would again hit the rotating mirror and emerge in a different direction to a detector. The light that hit side A of the rotating hexagonal mirror would not be seen until it emerged from the mirror at B which would occur at a time determined by the speed of rotation of the hexagon. If the length of the light path was well known the speed of light could

---

* In using the term 'light' I am referring to the whole range of electromagnetic radiation of which visible light is only one small section. Electromagnetic radiation covers everything from low frequency radio waves, through the infrared, visible and ultraviolet radiation to x rays and gamma rays. All these are 'light' and travel at the speed of light.

easily and accurately be determined (see Figure 2.1). Michelson used a light path 1½ kilometres long and refined the method to give a value for the speed of light of 299,774,000 metres per second. As the best available value today is 299,792,500 metres per second, it is remarkable how sophisticated his method was and with what painstaking care he worked.

Michelson is more famous, in fact, for another optical experiment he performed with E.W. Morley, in which they demonstrated that the speed of light was independent of the direction in which it was measured, and hence independent of the motion of the light source. Light can be described as a wave motion, but for waves to propagate some medium must be there to do the waving. For sound waves, air is the propagating medium, but for light, which can travel through a vacuum, what medium carries the waves? Nineteenth century scientists suggested that a luminiferous ether pervaded the universe and was the medium that transported light. But if such an ether existed it would surely mean that absolute space existed. Michelson and Morley wanted to establish the existence of the ether by

Figure 2.1 Michelson's Method for measuring the velocity of light. *A light ray is reflected off a rotating prism and then passes to a parabolic mirror which sends the beam down a long light path to another parabolic mirror which sends the beam back, so that it eventually falls on the opposite side of the prism. The time the light takes to cover this path is the total path length divided by the speed of light. When that time interval equals the time the prism rotates from one face to another a continual view of the light beam can be seen through the sighting telescope. Knowing the prism rotation rate and the path length then gives one the speed of light.*

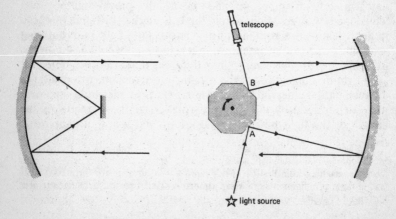

determining the passage of the earth through it. By comparing the speed of light in the direction of the earth's motion with its speed at right angles to that direction, the difference in the two values obtained would show how fast the earth was moving with respect to the ether. The experimenters expected that in one direction the speed of light would have added to it the velocity of the earth through the ether; and in the other direction it would have its 'standard' value. It was also expected that six months later the speed of the earth with respect to the ether would be subtracted from the speed of light, as the earth would be travelling in the opposite direction. But, in practice, the speed of light was the same value regardless of the direction of the earth's motion. The result of this experiment demonstrated that the ether did not exist, and that waves of light are really waves in a vacuum, with no medium propagating them.

The constancy of the speed of light and its independence from the motion of the light source or of the observer leads to further strange properties of time, as that quantity measured by a clock, which will be pursued in the next chapter. But before that I want to digress a moment to one last aspect of time measurement which concerns the smallest measureable intervals of time.

In a practical sense, and certainly in an operational sense, the smallest amount of time that can be measured would correspond to one cycle (or one frequency) of the transition in the caesium atoms of an atomic clock, that is to $\frac{1}{9192631770}$ of a second – about a nanosecond or 10^{-9} second. However, this does not mean that time intervals could not be smaller. The question is whether there is a lower limit to the size of a time interval; that is, does time come in basic units? Are there particles of time?

Scientists need to describe light in wavelike terms and in particle-like fashion in order to allow an explanation of the observed properties of light. Energy, too, has been quantized by physicists, that is, reduced to a basic packet size or particle or quantum of energy. The smallest amount of energy is called the Planck Constant after the nineteenth century German scientist, Max Planck, who first postulated the need for the quantization of energy. In recent years it has been suggested that space itself comes in discrete bits, that there are limits to how short a length can be, and that the unit, the Planck length, is 10^{-25} metres.

So if light, energy and space come in basic particles, why not time? The physicist David Finklestein coined the term 'chronon' for a particle of time and chronons are perfectly acceptable theoretical concepts. There are, however, some observations about chronons that should be mentioned. Firstly, there is no real need for us to have such entities. There are no observable properties of time that are not explained by our current, albeit vague, notions of time, so there is no motivation for the introduction of chronons into time theory, as there definitely was for energy and light. Then, there may well be no way of detecting chronons, for to do so would involve the timing of the duration of a chronon, and that would require a time interval shorter than a chronon, which is logically impossible. Neither is there any theory from which a prediction or a working proposal for the duration of a chronon could satisfactorily emerge. So the concept, although appealing, is rather *ad hoc*, and not of major or fundamental importance in present day physics. Nevertheless, it is interesting to consider what a chronon might be like.

The nanosecond is quite a long time. Light can travel about thirty centimetres in a nanosecond and thirty centimetres is an appreciable distance. In that length there are about a hundred million atoms, side by side. The speed of light, or the distance it travels in an interval of time, could be used as a way of defining the chronon. One might choose the time it takes light to cross from one side of an atom to another, which is about 10^{-18} second, for a chronon duration, but inside the atom is a central nucleus about one ten thousandth the size of the atom as a whole. Maybe the time it takes light to cross that should be a chronon of time, about 10^{-23} second. Such a time interval is inconceivably small, a thousand million million times shorter than the measureable nanosecond, and yet there are theoretical physicists for whom such a time duration is far too long. Some cosmologists, whose speculative research is to investigate the origin of the universe itself, would like to be able to handle time intervals of about 10^{-50} second, and still be able to distinguish separable events. So is the chronon really that incredibly small or is it just a wild idea, a meaningless description that says nothing about the nature of time at all? A wave/particle duality of time is a nice description, with time's periodic, cyclic flow neatly analogous with a wave model of the phenomenon of time, and with the graininess of instants of time nicely modelled by chronons – but this seems to be a description of reality rather than reality itself, and in this case, a rather fanciful one at that. Maybe time can be thought of as coming in chronons, but as of now, there is no fundamental theoretical need for chronons and certainly no way of measuring or detecting such particles of time.

This chapter has dealt with several aspects of time measurement

THE MEASUREMENT OF TIME

and the consequences that such measurements have imposed on a view of time. Time and motion are inextricably involved with each other; as are time and the speed of light. The next chapter moves on to a discussion of Einstein's Theory of Relativity and the strange properties of time that develop from it.

3

THE RELATIVITY OF TIME

What would the world look like if I rode on a beam of light? Albert Einstein pondered this question as a youth of sixteen and his answer – the Special Theory of Relativity – eventually changed mankind's way of thinking about the world, nature and time.

Einstein's theory arose from his attempt to reconcile Newton's Laws of Motion, which had stood the test of time for over 200 years, with James Clerk Maxwell's Laws of Electromagnetism, which had demonstrated in the 1850s that theoretically light was one part of a whole spectrum of radiation, from low frequency radio waves, through the infrared, visible and ultraviolet, to the high frequency x rays and gamma rays. Heinrich Hertz had demonstrated the truth in the theory by transmitting radio waves over a short distance, but one aspect of Maxwell's theory led to an inconsistency. The speed of light, and of all electromagnetic waves, was given as a constant by Maxwell's equations, and this speed and the existence of the waves themselves was independent of any outside effect. However, Einstein realized that if an observer was travelling alongside a light wave at the same speed as the light the wave would essentially disappear, as no wave peaks or troughs would pass by the observer. But the disappearance of light waves because of the motion of an observer should not happen according to Maxwell, so Einstein concluded that either Maxwell's equations were wrong or that no observer could move at the speed of light. He preferred the latter explanation for a particular reason.

The idea of relativity did not originate with Einstein, but goes back, certainly as a physical principle, to Newton and Galileo. Indeed, it was Galileo who first suggested that if someone was travelling in a boat on a calm sea he would have no way of detecting his motion except by looking out of the boat. Uniform motion cannot be discerned internally but only by reference to an external object. Newton's laws of motion, then, concern relative motion because in describing movement the reference always needs to be stated. If I am walking down the corridor toward the rear of a train at one metre per second and the train is travelling at thirty metres per second with respect to the railway track, my speed is one metre per second or twenty-nine metres per second depending on whether I regard the train or the surrounding countryside as a reference frame. So although Newton was concerned with absolute space and time, in practice he made use of this principle of relativity. One consequence of this principle is that the equations describing motion in one reference frame can be transformed into another reference frame by using simple mathematical expressions called Galilean transforms. Although Newton realized the difficulty in detecting the absolute reference frame which actually would be at complete rest, he nevertheless believed it existed. Einstein did not, so he preferred to argue that although in Newtonian physics a body could be accelerated up to and beyond the speed of light, this in practice disobeyed the principle of relativity.

The example Einstein used to illustrate this point was to imagine someone looking at themselves in a mirror, with the illumination coming from a stationary source behind the person. The person and the mirror are then accelerated up to the speed of light at which point the person/mirror system is moving at the same speed as the illumination from the light source. The light waves will never pass the person to reach the mirror and thus the image in the mirror will disappear. Einstein thought this was nonsensical and showed that it broke the principle of relativity, because it distinguishes one speed of motion as being absolutely detectable. He therefore concluded that Newtonian physics did not hold at speeds approaching the speed of light and used this argument in developing the Special Theory of Relativity.

From these preliminary remarks it is plain that the speed of light plays an important role in this theory, and so does the process of measuring time. Indeed, we can say, echoing Einstein's own words, that the Special Theory of Relativity arises from considering the measurement of speed and of time.

In order to obey the principle of relativity, and to preserve causality, Einstein made the assumption that any observer would always measure the speed of light to be the same constant value regardless of his own motion or that of the source of light. This may seem strange, and cases where it would appear ridiculous immediately come to mind. For example, consider a spaceship travelling towards a star at half the speed of light. It would seem natural to anticipate that experimenters aboard would measure the speed of

light from the star to be half that which they measured when they
were stationary. But, remember, Einstein said that there is no way
of knowing when a system is stationary absolutely as all measurements are relative to the person doing the measurement. As has
already been demonstrated, the speed of light can only be measured
by referring to a local clock because other supposedly synchronized
clocks are not reliable. In other words, an observer cannot measure
absolute space or time, but can only make measurements with
respect to himself and an arbitrarily chosen reference frame. So it is
not so surprising that the speed of light will be the same whoever
measures it. In returning to the spacemen above, it should be asked
how will they measure the speed of the light from the star? They
cannot time the passage of light from the star to them, because there
are no synchronized clocks on the star. The only practical measurement they can make is to divert the light as it passes them and then
reflect the same light back to their detectors (see Figure 3.1). In
practice they can only perform some version of the Michelson-
Morley experiment. This means that they have only measured the
speed of light with respect to themselves and hence cannot make
any claims about the speed of light relative to the star. The question
'what then is the *real* speed of the light?' turns out to be

Figure 3.1   *The way in which the speed of light could be measured by a
travelling spaceship.*

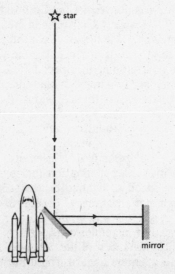

meaningless; furthermore, the only way to measure the speed of light relative to the star is to put someone on the star to measure it. That observer will perform a similar measurement to that of the spacemen and will measure the speed of light to be the same constant value that the space travellers measured it to be. Light, then, always travels at a constant speed with respect to the observer, not with respect to an imaginary ether, so light cannot itself be used to establish an absolute space or time.

By 1905 Einstein had formulated his Special Theory and answered his own youthful question about the appearance of the world if one could ride on a beam of light. He had shown that one could only perform such a feat in the imagination, but the answer one would discover even then was that time would stop. Light operates outside time. Light brings the information about the position of the hands on a clock. The information carried by the light doesn't change during its transmission. If the light leaves the clock at six o'clock, it doesn't matter how long it takes for the light to reach the observer, for when it arrives it is still saying that it left the clock at six. To that light beam it will always be six o'clock. To it, time will never flow. Its starting-off time will be the same as its arrival time. Light doesn't age.

The timelessness of light and its constant speed with respect to any observer lie at the root of Einstein's relativity theory because light is used to communicate across space and to make measurements of motion. Einstein regarded '... space and time as products of measurement', and relativity grew out of the measurement of time and the consequences of its communication between observers.

Before examining the details of Special Relativity, there is another property of time which is useful to examine. This is the property of the route dependence of time. Most quantities can be classified as being route dependent or route independent, by which I mean whether their value depends on the path chosen in making the measurement. For example, two towns might be separated by forty miles 'as the crow flies', and that distance is route independent. However the physical separation of those two towns is measured, the distance between them will be forty miles. Now imagine there are two roads joining those towns, one of which is narrow, but reasonably straight, and which passes through several small villages, and the other road is a wide, sweeping motorway, but which makes quite a detour to avoid the villages, in passing from one town to the other. The mileage reading on a car driving from one town to the other may be forty-five miles on the narrow road and fifty-five miles on the motorway. As fuel consumption is primarily a measure of distance travelled less fuel will be used on the shorter route and fuel consumption is therefore a route dependent quantity. The journey time, too, is route dependent, as it takes an hour and a quarter to drive on the narrow road, but only forty-five minutes on the motorway going at seventy miles per hour.

Space and time intuitively seem to be route independent quantities. The actual separations of places or times do not seem to depend on the route followed. However, the consequence of the analysis that time is relative to the observer, because of the limitations of what can actually be measured imposed by the passage of light, means that time is route dependent. The above example about a car journey from one town to another shows that the journey time was route dependent. This is quite obvious and can be explained by the speed at which the journey was undertaken. Route dependence of time is a relative experience and at the same time it can occur in an 'absolute' reference frame in which the separation of the two towns can be determined 'as the crow flies', because the observer is stationary within this reference frame. Leaving aside the local vicinity in considering larger scale distances and speeds, the situation becomes more complex. For a start, a spaceship cannot determine whether or not it is at rest in the reference frame of, say, two stars it wishes to travel between, so the actual separation of the stars may not be quite so easy to determine. This means that the time taken for a journey, which is route dependent, may not be the same for the traveller and for someone watching him! Time is dependent on the speed of travel.

This rather curious observation can be approached from another vantage point by returning to a discussion of simultaneity. In the last chapter it was made clear that two events are simultaneous when they occur at the same moment in the history of a clock, but if events are widely separated in space then their simultaneity is harder to establish. Einstein, again arriving at a conclusion by considering the interaction of the measurements required and the finite but constant speed of light, showed that simultaneity was also affected by the motion of the observer.

In order to be free from the complications of earth-bound systems an example involving space travellers and their observations of stars will be used. Imagine Spaceship A is travelling quickly along and its pilot is watching two nearby satellite beacons (the space age equivalent to buoys at sea). His radar scanner picks up two approaching spaceships, one coming from the front and the other approaching from the rear. They are both identical distances away and travelling toward him at identical speeds. It doesn't take him long to realize that they will both pass him by at the same moment. Now the paths of these spaceships are such that they all three pass at the same instant and a moment later the pilot in Spaceship A notices that both beacons simultaneously light up.

Later that night (if you can determine day from night in interstellar space) the pilots of the three spaceships all meet up in their neighbourhood space bar. Pilot A says, 'Hey, did you notice those two beacons switch on simultaneously when you both passed me today?'

'Well, yes', says Pilot B (he was the one approaching A from the right), 'except that the left hand beacon came on before the right hand one'.

'No, no', interrupts Pilot C, 'the right hand beacon lit up first, I noticed that particularly.'

In what order did the beacons really light up? Who is correct? The answer, unfortunately, is that all three pilots are correct. Simultaneity does not mean the same thing when dealing with moving observers and separated events. Figure 3.2 shows what happened. Regard Spaceship A as being the reference frame for this example and therefore effectively stationary; it must also be stipulated that the two beacons are stationary within A's reference frame, that is they are drifting along at the same velocity as A. Now at the instant that B and C pass A (Figure 3.2) the two beacons light up. As they are a certain distance away (let us say they are 300,000 kilometres away) it will take the light from them a finite time to travel to A. They are at such a distance that light takes one second to reach A from both beacons, so one second after they switched on A sees them light up simultaneously. The fact that he does see them switch on together means that in effect (if not in practice) the two beacons are each an equivalent distance from A so the light paths seem equal.

In the second that it takes the light to travel to A the other spaceships have moved on a considerable distance as they are travelling at high speed with respect to A. This means that by the time light reaches them the light path from one of the beacons will be greater than from the other, and the situation will be effectively the same for each space craft, except that each will see a different beacon as 'being nearer'. This difference in light path means they see the beacons switch on in a different order, and hence the disagreement between the three pilots in the bar that night.

a.

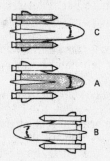

b.

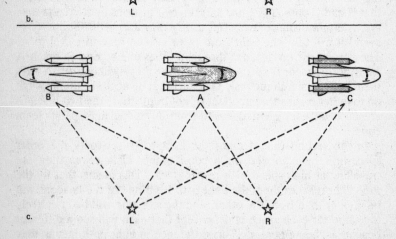

c.

Figure 3.2   *Illustration of the relativity of simultaneity.*

This analysis looks contrived because it can always be said that the beacons did switch on simultaneously as they were at rest in A's reference frame. But this is a false argument because there is no way of knowing the 'true' sequence of the beacons lighting up except by measurement and all three observers measure a different effect. The purely arbitrary basis of observing this phenomenon from A's reference frame is deceptive. A more valid observational condition is to say that the three pilots saw the beacons switch on in three different ways. The initial conditions in this example can be relaxed by saying that nothing is known of the motions of the two beacons except that when switched on the observers saw them in the order already described which makes for a more general situation and clearly demonstrates that simultaneity is a quantity entirely dependent on each observer and his motion relative to the effect. It is an example of the route dependency of time and becomes increasingly larger an effect the faster observers move with relation to each other and the greater the separation of the events whose simultaneity is being observed. Naturally, when events and observer are relatively at rest and the two events are not separated in space then events can certainly be called simultaneous, but this is a special case, even though it seems the norm. What the example has shown is that this definition of simultaneity – the concurrence of events at the same moment in the history of a clock – is still true in wider application and that it leads to the apparent oddity of the relativity of simultaneity.

Einstein's Special Theory of Relativity is concerned with the description of movement and how objects appear when they move relative to each other. As the Theory is a special case of a more general theory, it limits itself to descriptions of objects in uniform relative motion. Einstein's arguments were based on the two postulates he proposed, namely, that absolute motion was undetectable and that the speed of light was constant for all observers. Some of the consequences of adopting these postulates have already been demonstrated. There is also a prediction about time based on those postulates which forms part of the theory.

Einstein showed that time as perceived by an observer appears to flow more slowly in systems in relative motion to himself. Furthermore, the faster the other system was moving, relatively, the slower time would appear to run in that system, so that if that system moved at the speed of light, time in it would stop still. Mathematically, the time in a moving system, as seen by an observer, and compared with his time, would be slowed down by the factor

$$\frac{1}{\sqrt{1 - \left(\frac{v}{c}\right)^2}}$$

where v is the relative speed of the moving system and c is the speed of light. One can immediately see that if v is zero then the two time systems are identical, and that if $v = c$ then the time dilation would be infinite, i.e. time in a system moving at the speed of light is nonexistent. At intermediate speeds the effect is something in between these two extremes but because we are dealing with the square of the ratio v/c and the square root of the difference between that ratio and unity, the effect is rather small until v becomes quite large. This can be seen in Figure 3.3, where the time distortion factor

$$\left(\frac{1}{\sqrt{1 - \left(\frac{v}{c}\right)^2}}\right)$$

is plotted against the fraction of the speed of light (v/c) of a relatively moving object. It can be seen that even at a speed of 30,000 kilometres per second (one tenth of the speed of light) the size of the time dilation effect is less than one percent different from a stationary clock and the effect is only appreciable at much higher speeds. However, remember that the speeds and the times are all relative to the observer in the stationary reference frame. If I was in a 'stationary' space station and watched a spaceship travelling rapidly away from me I would see his clock running slower than mine, but the spaceman would see my clock running slower than his because the effect is completely symmetrical. To the spaceman I would be moving away from him and my time would be dilated. This is because time is relative to the observer, and the only thing discerned is what the observer 'sees', not what absolutely is.

This 'time dilation' effect is most clearly illustrated on a space/time diagram, where distance travelled is drawn horizontally and time vertically. In Figure 3.4 the speed of light is indicated by the 45° line that travels one unit of distance in one unit of time. This is called the light line and as nothing can travel faster than light nothing in the diagram can have a slope less steep than the light line. If I regard myself as stationary I will move up the vertical axis, not travelling any distance but moving forward in time. The spaceship

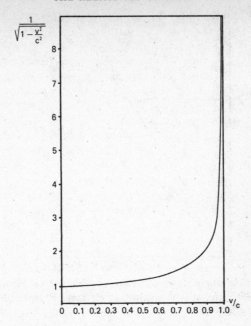

Figure 3.3 *The time dilation factor*

$$\left(\frac{1}{\sqrt{1-\left(\frac{v}{c}\right)^2}}\right)$$

plotted against the fraction of the speed of light ($^v/_c$) showing that departures from 'normal' conditions do not become significant until very high speeds are reached.

travelling away from me is moving at a high relative speed, say about seventy-five percent the speed of light. As I watch the spaceship move away from me, one hour on my clock would be matched by only 40 minutes passing on the moving clock, as I see it. So if I wanted to see an hour pass on the travelling clock I would have to watch it for about an hour and a half as indicated on my own clock. This dilation effect is simply given by the above factor. So at seventy-five percent of the speed of light the time dilation factor is about one and a half times the time of a relatively stationary clock, which is indicated by the length of the line in Figure 3.4.

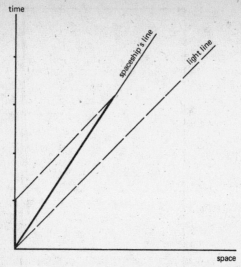

Figure 3.4 *Space/time diagram showing the light line and the world line for a spaceship travelling at three-quarters of the speed of light, illustrating the time dilation effect.*

Figure 3.5 *Diagram illustrating time dilation.*

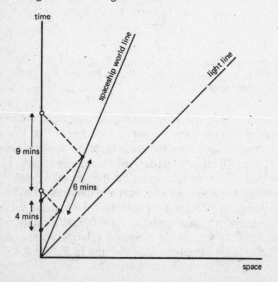

If I wanted to check time intervals with that spaceship I could send a flash of light to it at, say, four minute intervals, and prearrange that whenever the spaceman saw a flash of light from me he would send a flash of light back. The space/time diagram of Figure 3.5 shows the result of such an experiment. My light travels out at 45° (the speed of light) and the spaceship receives the 'flashes' every six minutes, because it has travelled so much further by the time the second flash has reached it. The returned flashes travel back at the 45° angle on the diagram and so I receive those signals separated by nine minutes. Four minutes of my time has been stretched out to nine minutes in the spaceship, as I see it. Of course, on switching to the reference frame of the spaceship precisely the opposite effect would be noted because the situation is time symmetrical. If instead of light flashes I sent an image of my clock and received back images of the spaceship's clock, the space pilot would see my clock had advanced four minutes between flashes, but he would have timed these over six minutes, concluding that my clock was running slow. If he sent independent images to me every four minutes, I would receive these every six minutes on my clock. However, I would now like to advance to a case involving a third spacecraft in which the effect no longer remains symmetrical.

Imagine a spacestation which dispatches Spaceship A off to relieve an outlying station. The spacecraft travels away from the spacestation at a uniform speed of about 225,000 kilometres per second in a straight line directly to the outlying station and the instant it arrives the crew, who were waiting to return, set off straight back on Spaceship B at the same uniform speed that the outgoing craft took. (I shall ignore the speeding up and slowing down procedure for simplicity). The pilots of both the spaceships are asked to flash a signal to the home station every four minutes by their clocks which have been checked as thoroughly as possible to ensure that they are identical to those at the spacestation. At 10:00 a.m. Spaceship A sets off, flashing its light every four minutes and does so ten times, so that when it reaches the outlying station its clock reads 10:40. At that instant the pilot in Spaceship B sets his clock to 10:40 and comes speeding back to the space station, flashing his light ten times, every four minutes. As he arrives back, his clock reads 11:20, just as would be expected.

However, what does the home station's clock read? The ten flashes sent by the outward travelling craft reached it every six minutes, because in the passing interval the craft had gone so much farther away that the light took longer to reach the station. So

Spaceship A is 'seen' to have reached the other space station at 11:00 a.m. by the home station clock. The conclusion is that because of the relative motion, time on Spaceship A has been slowed down. At 11:00 on this same clock Spaceship B sets off home and the station sees flashes from its light every 2.7 minutes, rather than every four, because between flashes the craft has moved a lot nearer and the light has less far to travel. So after the ten flashes have reached home base, and indeed so has Spaceship B, the clock reads 11:27 a.m. If this clock is compared with the one on board Spaceship B, it is found to be seven minutes ahead. Time has passed at a different rate for the home station crew than for the travellers.

Because there are three participants in this story there is no symmetry between one observer and the other two. The situation is shown in Figure 3.6 and however the reference frame is changed the same result always occurs. This is because in the story the spacestation's reference frame is unique. In its frame one craft travels away from it and the other approaches. In A's reference frame both the station and Spaceship B travel away from A and in B's reference frame the station and Spaceship A approach B. So time for the 'moving' spacecraft actually does flow at a slower rate than for the stationary object; relativity is no longer purely relative!

This result might seem to be a clever mathematical trick which doesn't relate to reality at all, and certainly no such experience exists in everyday life. And yet the effect does occur and has been experimentally demonstrated. In 1972 an atomic clock was flown around the world in a jet airliner and it did age less than its stay-at-home twin in the laboratory. The difference in time between the two clocks was only measureable at the extremely short time intervals with which atomic clocks operate, but the result matched Einstein's prediction. Time dilation does occur.

The above imaginary example involved only uniform motion in a straight line, which is the restriction imposed by the Special Theory of Relativity. The atomic clocks flown around the world involved changes of speed and changes of direction and therefore accelerations, and the General Theory deals with those cases. But before examining what General Relativity has to say about time, let me just mention that two other effects predicted by Special Relativity are length contraction and mass increase. Objects moving at high speeds relative to an observer appear to become shorter in length and their mass, that is, their resistance to change of speed, becomes greater. Einstein predicted that if an object could reach the speed of

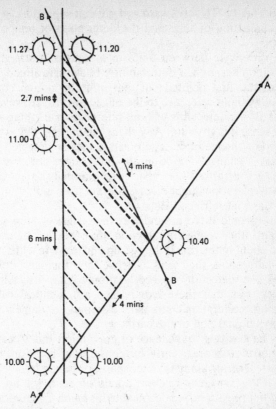

Figure 3.6 *Diagram showing the different rates at which clocks run in systems in relative motion, as described in the text.*

light then its length would reduce to nothing and its mass would become infinite. As such effects are impossible, the conclusion is that the speed of light is unobtainable, hence preserving the principle of relativity as given by Galileo and Newton and Maxwell's description of electromagnetic radiation – but only at the expense of discovering that time and space are not exactly what they appear to be. So far, the application of this approach has been pushed almost to its limit.

Relaxing the special condition of uniform motion in a straight line to allow for acceleration and for a General Theory of Relativity turned out to be a very much more complex matter than might at first appear, and Einstein's General Theory is not at all like the

Special Theory. This awkward and difficult subject has a bearing on this examination of time and the description of temporal characteristics.

The difference between uniform motion in a straight line and acceleration lies in their detectability. I have said already that it was Galileo who first pointed out that uniform motion could not be detected without reference to the outside world. However, acceleration is always detectable without reference to anything outside and in that sense it is absolute. Any change in motion, that is any change in speed or change of direction, can always be detected absolutely. Knowing a change in motion absolutely does not reveal absolute speed, of course, but the capacity to internally detect that change makes acceleration qualitatively different from uniform motion, and it was this that intrigued Einstein.

When people drive around a corner, start or stop suddenly or make any other change in direction, they feel the change. They get thrown about in the car. If they go up and down in a lift, or on a ride in the fairground, they tend to 'leave their tummies' behind them. This is the detection of acceleration. What Einstein realized, however, was that these experiences are identical with the all-pervading acceleration continually experienced, namely that due to the gravitational pull of the earth.

Why he saw this equivalence of gravitation and acceleration can be understood more clearly by considering a spaceship with no windows. Drifting along in outer space, or falling freely towards the earth, if the spaceman releases a held object it will not fall to the ground but remain suspended next to his hand. The reason for this is that the whole system and everything in it shares the same motion and will continue to do so while uniform conditions continue. It is not, nowadays, uncommon to see films of a man in space letting go of a hand held object and seeing it float in midair, or of seeing free falling parachutists doing similar tricks. Locally, these systems are free of gravitational attraction and so uniform motion is possible.

If a ball were rolled along a table and permitted to continue past the table's edge in such a free falling system it would continue in a straight line across the room. This is true for all 'particle' paths across or through any free falling system, including light beams; they will all be straight lines, at least in the local reference frame. The 'normal' laws of physics, by which I mean our terrestrial experiences of physics, do not seem to operate properly in such a laboratory. For example, a pendulum will not swing in a free falling system. But of course Newton's Laws of Motion operate perfectly in such a setting, far better than on earth where the acceleration due to gravity is always in operation.

If the windowless spaceship fires its rockets so that it accelerates forward things behave differently from when it is in uniform motion. A hand held object will fall to the floor and pendulums will start swinging; crew members will experience their weight again and not float around the room. The reason for this is that the craft is always moving faster each moment and so the floor is always coming up toward the object released, so the object is seen as accelerating toward the floor, although it really is only moving with the uniform motion it had on release. When a ball is rolled along the table it will continue past the table's edge but will follow a parabolic path toward the floor, just as it would on earth. Again, this is because the floor is accelerating upwards. In the accelerating reference frame the laws of physics appear quite 'normal', that is, terrestrial, and particle paths are observed to be curved. If a light ray passes through the accelerating system its path, too, will be curved for the same reason that the path of the ball is. Of course, to be able to detect the curvature of the light beam one must be accelerating highly, but detectability is a practical problem only.

Einstein said that the people in the accelerating spaceship would not be able to tell whether the phenomena they experienced were due to the firing of the rocket engines to cause the acceleration, or whether the rocket was stationary on a planet with a certain gravitational pull. The physics is terrestrial in the accelerating system, and it is so because of the equivalence of gravitation and acceleration. This principle of equivalence is the basis for Einstein's General Theory of Relativity.

As gravitation is the force that binds the whole universe together, the force which operates between the stars and galaxies, the universe could equally well be described as an accelerating universe. The physics of the universe could be expected to be like the physics in the accelerating space ship; that is, pendulums would swing and particle paths would be curved. Also, light paths should be curved. Indeed, the first prediction that arose from the General Theory was that the light of distant stars would be 'bent' around the sun, as shown in Figure 3.7. A star seen from earth as near to the sun, Einstein explained, has a real position closer still to the sun. As stars in the neighbourhood of the sun can only be seen during a solar eclipse there are few occasions to demonstrate this effect, but it was first confirmed observationally by Eddington in 1917 and proclaimed a triumph for Einstein's theory.

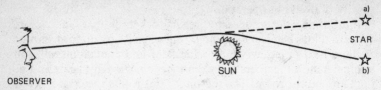

Figure 3.7 *The gravitational bending of light, illustrated for a star that appears at position (a) when seen at a solar eclipse, but which is really at position (b) when its light path has not been bent by the gravitational mass of the sun.*

As light paths must be, in general, curved because of the gravitational mass of the universe, then the fabric of space and time must also be curved, argued Einstein. Intuitively, space feels linear, and light rays seem to travel in straight lines, but General Relativity demonstrates that what appears straight is really curved and so instead of a 'flat' universe it is found to be a curved one, just as the flat earth turned out to have curved geometry. Just as on earth the shortest distance between two points is the curve of a great circle and not a 'straight' line, so too in space a curved path, or light line, is the shortest distance between two places. Locally it appears straight, just as the earth's surface appears flat, because the curvature is small on a local level. That explains curved space, but what about time?

Again, consider two spaceships, one of which is moving along in a straight line at a uniform speed and another one which is accelerating away. On a space/time diagram such as Figure 3.8 the path of the accelerating craft will be a curved line because the rate of change of distance with time is increasing. If the uniformly moving spaceship sends out signals every four minutes as before and the accelerating ship flashes its light when it receives the signal then the rate at which time passes on the two craft can be compared. As can be seen in the diagram, the path length of the accelerating ship's line increases between each successive flash sent out by the uniformly moving ship, and the interval of time passing before the reciprocal signal is received back increases the greater the acceleration. Hence the first four minutes in the 'stationary' spaceship are dilated into five and a half minutes on the accelerating vehicle, and the second four minutes are stretched out to over six minutes, so that the return signals are then received back at seven and twelve minute intervals. The conclusion of the observers in the uniformly moving craft is that

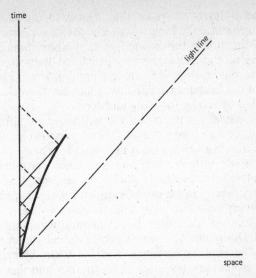

Figure 3.8 *Space/time diagram for an accelerating system, showing why the relatively stationary observer sees the clock in the accelerating frame running ever more slowly.*

the greater the acceleration of the other ship the greater the time dilation effect, and that time in the accelerating ship passes more slowly. But there is no symmetry between the two vessels any longer.

In the accelerating ship the crew expect to receive signals at equal intervals and so as the flashes arrive they count off the minutes. As more time passes between each signal from the stationary ship they conclude that time is flowing faster in that vessel. Or, conversely, they could conclude that their acceleration has actually slowed their own clocks. All accelerations have a physical effect on clock rates, whether the clock is gravitational, atomic or biological.

The effect of acceleration is to slow the flow of time relative to a non-accelerating system and this effect is absolute. Time, in a sense, is curved like space, and an accelerating system follows a curved path while uniformly moving systems follow straight paths. The two types of motion follow different routes from one place to another and time, I have already pointed out, is a route dependent quantity.

This asymmetry between the accelerating and non-accelerating systems was shown in the previous example where the uniformly moving spaceships going to and coming back from another space station had different times on their clocks to the 'stay-at-home' clock (for reference see Figure 3.6). The asymmetry was due to the change of direction in the outward going and homebound journeys, and although the individual motions remained uniform, that change of direction was equivalent to an acceleration. It was also a simple example of the famous twin paradox, which becomes clear once the time dilation of acceleration is understood.

The twin paradox is not actually a paradox at all, although when first postulated appeared to be so. One twin stays at home while the other twin travels away in a spaceship, returning home eventually to discover his stay-at-home twin has aged much more than he. The example already discussed (Figure 3.6) demonstrates this effect quite clearly. The paradox can be explained using special relativity, but it appears paradoxical in that framework because the time dilation effect is relative, and should be quite symmetrical between the twins. The reason it is not is because acceleration must enter into the situation if the travelling twin is to return home; that is, he must turn around at some stage in his journey and that constitutes an acceleration and immediately introduces the asymmetry.

Imagine that a fifty year old man and his twenty-five year old daughter each have their twin atomic clocks. The father and his clock get in a rocket which accelerates at ten metres per second every second. This acceleration is equal to the acceleration due to gravity on earth, so this travelling man feels quite at home in his rocket with what appears to him to be terrestrial conditions. He keeps his rocket engine firing for a month, by which time he will have gained a speed, with respect to the earth, of close to ninety percent of the speed of light. At the end of the month the rocket is turned around, subjecting the father to a little momentary discomfort, but the engines are kept blazing and the craft decelerates until at the end of the second month it is at rest with respect to the earth. The rockets keep burning and the craft accelerates back towards home for another month, when again it is travelling at about nine-tenths of the speed of light. Again, it is reversed so that for the final month of the journey it is slowing down and lands back on earth. The father checks his atomic clock and assures himself that his journey has taken him exactly four months. He opens up the door of the rocket and there is an elderly woman to greet him.

'Father,' she cries, 'how wonderful to see you.' The man's daughter is now sixty years old, ten years older than her father, for time ran more slowly in the rocket than it did on earth, and the stay-at-home atomic clock has ticked its way through thirty-five years. As

far as the daughter was concerned the father travelled to a distance
of about fifteen light years away and didn't send a signal that he was
on his way home until she was in her fifties.

The story may sound fantastic but it follows the same principle as
the earlier example, where the time difference was only a few
minutes. Of course, this case involves sustained accelerations that
could not possibly at present be produced technically, but this effect
can be measured by flying atomic clocks around the world in
airplanes. The effect can also be verified with radioactive particles
in the giant particle accelerators used by physicists. And scientists
have even calculated the size of the time dilation for the astronauts
who went to the moon in 1969. On their return they were five
thousand millionths of a second younger than their wives who
stayed at home.

In fact, the twin paradox is largely an example of the effects of
Special Relativity, and the only part played in the story that in any
sense involves the General Theory is the asymmetry between the
two observers introduced by the acceleration (as time dilation itself
is not the effect of General Relativity). However, Einstein showed
in his General Theory that acceleration does slow down the flow of
time and as acceleration and gravitation are equivalent he predicted
that clocks would run more slowly in a strong gravitational field.
This means, for example, that a clock in outer space will run at a
different rate from one on earth, and a terrestrial clock will run
faster than one on a more massive body like the sun.

The strength of a gravitational field depends partly on the mass of
the object giving rise to the field and also on its size. The more
massive the object the stronger the gravitational field; and clocks on
that object will run more slowly. Also, for a given mass, the smaller
the object is, the more condensed its mass will be, and hence the
stronger the gravitational field. Related to this last factor is the fact
that the field becomes stronger as the centre of the massive object is
approached. This implies that the force due to gravity on earth will
be stronger at the bottom of a mountain than at the top, and thus
that time flows more slowly at sea level than it does in the hills. This
remarkable idea has, in fact, been confirmed by experiment, again
by using atomic clocks. A clock at the bottom of a tower does run
more slowly than one at the top. The effect is incredibly small
(about one part in a thousand, million, million) but measureable.

Atoms behave like clocks, as discussed in Chapter Two, because
they absorb and emit light of specific frequency, that is, time
interval, and Einstein therefore predicted that the light coming from

atoms in a stronger gravitational field than on earth would have lowered frequencies. For example, the atoms of the sun would emit light at frequencies two parts in a million lower than in a terrestrial source. The observation of this difference is not too difficult to make and has been amply confirmed, indicating that clocks on the sun actually run slower than clocks on earth. To observe this gravitational time dilation on a larger scale a neutron star provides an example of a source of strong gravitational field.

Neutron stars are stars that once were not unlike the sun, except that they were a little more massive, but at the end of their life were unable to sustain the processes of energy generation which keep stars burning and radiating light. Stars such as the sun maintain their size by balancing their natural inclination to collapse and condense due to their strong gravitational pull with an outward pressure from the energy generating processes deep in their interiors. At the end of a star's lifetime the fires go out and the star collapses under its own weight. Most stars collapse down to a size about that of the earth and slowly cool off like a piece of glowing coal. However, more massive stars collapse still further because their weight actually disrupts the individual atoms of their material. When the atoms themselves collapse the particles comprising the atoms coalesce to form neutrons and the whole star shrinks down in size to about twenty kilometres across, not much bigger than the size of a large city. Such an object is called a neutron star and its gravitational field is almost a million times that of the sun. Time flows very much more slowly on such a star. Light can only leave the stellar surface by giving up a large proportion of its energy and thus its wavelength increases and its frequency decreases. This altered time interval indicates that time flows more slowly on a neutron star.

Just as an acceleration can be shown on a space/time diagram as a curved line, so a gravitational field curves the space/time around it so that light near such a source follows a curved path rather than a straight one. The greater the curvature of the light path the slower time flows. A black hole is produced when the curvature is so great that light curls back on itself – and time stops still.

If a collapsing star is more massive than that which resulted in a neutron star, there reaches a point at which the neutrons themselves cannot resist the overwhelming force of the infalling weight of stellar material. On collapsing even further into itself, the star passes through a stage at which light emitted from its surface cannot any longer escape because the gravitational field is so strong that the light curves round and falls back to the star. When that stage is reached the star has collapsed beyond its 'event horizon' and can no longer be seen. It has become a black hole. Time, in the black hole, has ceased, and time, as the event horizon is approached, runs slower and slower and stops when the horizon is reached. The black hole, in some respects, is the gravitational equivalent of the speed of light.

Of course, if black holes actually exist, and so far one has not been positively observed, then, unlike the speed of light, a travelling astronaut could visit one. He would not be able to return from the visit, however, and once he had passed across the event horizon he would not be able to communicate with an external observer. As he travelled towards the black hole his signals, the usual flashes of light, would come at ever increasing intervals and his image, as he reached the event horizon, would remain for an infinite time although he would pass across this imaginary barrier unaware of it. Once inside the black hole, however, space and time would alter. The spaceman would no longer be able to move about freely, but could only move along a path toward the black hole's centre. However, he would be free to move about in time! In practice the gravitational forces would physically overwhelm him and he would rapidly be crushed out of existence, but if the black hole was big enough so that the tidal forces would not overwhelm a visitor, and especially if it was rotating, then life inside the black hole would be very interesting.

As time does not exist in the black hole the astronaut could travel in time, thus meeting himself on his journey as many times as he wished. He might find that the black hole actually led him into another part of the universe. It would have to be so distant that his arrival there could not causally affect the place he had come from, but he could settle down to a new life there. Alternatively the black hole could lead him into another universe altogether. This other universe is indescribable by the laws of physics simply because these laws cease to operate in the black hole, at least to the extent of predicting the properties 'on the other side'. If the astronaut didn't like this new universe he could step back into the black hole, which he notices he shares identically with this alternative universe, and enter yet another universe. He could go on playing this game until he got really bored, but he could never re-enter his original universe.

Rotating black holes have different properties from non-rotating ones, and one such difference is that they possess an ergosphere, a region outside the event horizon that shares some of the properties with the black hole and from which work can be extracted. An object can enter the ergosphere and leave it again, but only at the expense of allowing half its mass to fall into the black hole proper. The energy the remaining half brings out is paid for by a reduction in the rate of rotation of the black hole. This is a useful property because, for example, a dual stage rocket could be fired into the ergosphere so that half of it, previously loaded with refuse, could fall into the black hole, and the part that could be re-used could re-emerge, and re-emerge with more energy than had originally been put into it! One difficulty, however, is that the space and time bearings of things in black holes cease to be significant and so the ejected rocket could turn up anywhere or at any time. Travel out of a black hole, at least out of the ergosphere, is undeterminable in either space or time.

I have come a long way from operational definitions of time and the experiments to determine the speed of light. Yet what I have postulated about time and black holes is one end point in a long line of scientific enquiry based on the idea that time is that quantity measured by a clock and that for different observers to say anything meaningful about time at any place other than where they are themselves can only be accomplished by exchanging signals, of communicating via electromagnetic radiation. The Special Theory of Relativity and those parts of General Relativity touched on are the extreme outcome of using that model of time. One of the ways in which that model has had to be pursued is by linking space and time into the four dimensional space/time continuum of which a space/time diagram is but a simplified representation. Such diagrams are useful, of course, and have helped to demonstrate some of the consequences of the strange behaviour of time in relativity theory. Space/time as an idea was actually introduced by the mathematician Minkowski in 1908 and the space/time diagram of Figure 3.9 is called a Minkowski diagram.

Such a diagram consists of the 45° light line emerging from 'here and now' in both spatial directions and continuing back into the past and forward into the future. Such a diagram is static and highly spatialized but it illustrates the point that at the 'here and now' there is a past that can only have been causally connected to 'now' by events inside the light cone, and similarly anything done in the future will be confined to the region inside the light cone ahead of 'now'. This is because nothing can move faster than light and hence no causal influences can cross the light cone barrier. Events outside the light cone could be past or future in relation to 'here and now' and will depend on their relative motions and accelerations. Flashes of light from that region will reach the light cone at some time but no significant statements can be made about when or in what order they occurred. In a black hole everything is outside the light cone. The cone itself has been reduced to nothing. However, in the Minkowski diagram, there is no flow of time. It is a frozen moment of 'now' that is portrayed in the picture. To that extent it excludes such flows as consciousness, and only relates what is permissible materially and causally as constrained by the speed of light.

Figure 3.9 *A Minkowski diagram. This form of the space/time diagram shows the three dimensional nature of the light cone stretching into the past and future. An observer is always at here and now and can only have causally related access to events within the light cone.*

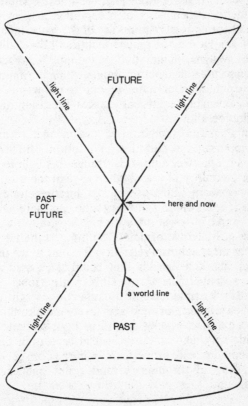

So, what have the ideas of relativity to say about time? Certainly it is an area of science that deals with time more explicitly than most, and the consequences of the theory have odd things to say about time. The route dependence of time has been encountered as well as time dilation and the idea that time is relative to the observer. Each person carries his own time and it will not necessarily be the same time as for other people. Also, time depends on one's relative motion, on one's acceleration and on where one is in the universe. As gravitational fields permeate the whole universe there is nowhere where time runs at a 'proper' rate; it runs slow everywhere, again reinforcing earlier ideas on the lack of an absolute universal time.

Apart from these things the theories of relativity have little else to say about time. The theories are, after all, only descriptions of those aspects of reality that are accessible, via a developed mathematical background, to certain types of observation and experiment. To discuss the theories at all requires use of the space/time continuum, which over-emphasizes the spatial qualities of time, and so, like all descriptive systems, limits understanding. However, relativity theory allows a great deal of exploration within the limitations of the operational definition of time. I hope to show in the next two chapters that even in the physical sciences there is more to time than such a definition allows.

Two things have emerged from this discussion of relativity which I regard as important and to which I will return. The first of these is the timeless property of light. It is strange that light to an observer has a finite measureable speed, when to light there is no time and therefore no space. Light seems to interpenetrate the whole universe as if everywhere and every time was 'here and now', and yet to the human observer it travels and it travels at a finite speed. That is one great puzzle of time. The other is the individuality of time that has emerged from relativity. It seems to me that although man moves about at speeds that could not possibly be called relativistic, he nevertheless carries his own personal, proper time with him. But for now I will leave relativity, which has emerged from the measurability of time and its communicability from one observer to another, in what has turned out to be an essentially deterministic way, and consider instead aspects of time that are related to indeterminate properties of nature which encompass both the gross scales of the universe and scales smaller than atoms themselves.

4

TIME FORWARD AND
TIME REVERSED

Time flows and it flows forward. It is neither observed nor experienced flowing backward. If it did, it would seem awkward, disturbing and unnatural.

In this and the following chapter some answers given by science to the questions of why time flows and why it flows in one direction will be examined. The odd thing about time flowing and flowing in one direction in relation to science is that in most of physics, time, as it enters the equations, neither flows nor flows asymmetrically. This total incongruity between the description of time as presented by physics and the human involvement and experience of it certainly indicates that something is missing in these equations.

The equations of the physical sciences show time as symmetrical, that is, the equations work as well in one temporal direction as in the other. It is like watching a film of the central, uniform patch of a billiard table and seeing balls rolling around and colliding (especially if the cue or the pockets are not visible). In such a case it would be impossible to tell whether the film was being run forward or backward. Simple mechanical collisions and motions can easily appear, as they are described in the equations, to be symmetrical in time. Indeed, much of physics consists of symmetries, but this book is not the place to explore that aspect of the description of nature. One major asymmetry in the world is certainly that of time, and since there are no clues to this asymmetry in mechanics and dynamics, another area of the physical sciences must be pursued. I want to look first for an answer in the field of thermodynamics, which will introduce statistical and probabilistic answers, and then pursue that approach in subatomic physics. The following chapter looks to cosmology and the temporal properties of the universe at large to provide another answer to the problem of the arrow of time.

Where does the flow of time become important enough to be taken into consideration in physics? The equations of mechanics and electromagnetism are time symmetrical but the experimenter, testing those equations in various applications uses a clock of some kind to measure time intervals. But time doesn't really enter the scene because the experimental clock is merely counting regular and periodic intervals of frequency. It matters not to the experimental physicist whether his clock is running forwards or backwards, providing the time intervals are consistent in either direction. Duration, too, is time symmetrical.

Periodically based clocks do not indicate, at least primarily, a flow of time. They merely add up successive intervals of pendulum swings or crystal oscillations. The only indication of flow is indirect, seen for example in the slowly falling weights of a pendulum clock or the running down of the batteries of a quartz watch. Such periodic clocks provide accurate measures of interval but do not directly indicate flow except that the duration of each single periodic interval represents a symmetrical passing of time indicated by the changing property of the oscillating mechanism. One clock, of course, that does indicate this quality is an hour glass. The falling particles of sand not only demonstrate a flowing action but also flow in only one direction. The grains of sand do not jump back into the upper reservoir, but spontaneously fall down and the irreversibility of this process says something about the arrow of time.

To elaborate, an imaginary clock consists of two reservoirs separated by a small aperture in the intermediate partition. The reservoir on the left is filled with a coloured gas, such as bromine vapour, the aperture is kept closed and the right hand reservoir is empty. To operate the clock the aperture is opened and the device observed as the gas flows through the aperture, by diffusion, until the whole of both reservoirs are uniform in density. The time taken for this equilibrium to be reached, which will depend on the ambient temperature, can be calibrated against a standard clock and hence provides a form of hour glass (although admittedly not a very useful one). The advantage of such a device over a conventional hour glass is that it depends for its operation only on the random motions of the gas particles and not on an external trigger such as the gravitational field of the earth.

The property this clock shares with the hour glass is that gas does flow through the aperture, although not as dramatically as the sand through the neck, but the flow is irreversible. And this irreversibility seems to be connected with the arrow of time. If two photographs – one showing the coloured gas filling only the left compartment and the other showing both tanks full – were taken of the air clock, it would be obvious which picture was taken first. Gas molecules spontaneously fill a volume and do not suddenly confine themselves to a small part of that volume (just as sand grains do not all jump back up to the top of the hour glass). But surely, it could be argued, the gas molecules could all flow back. Once that density equilibrium has been reached between the two chambers, molecules from both sides are crossing over and it could be imagined that at some moment all the molecules in one side could have exactly the right speeds and directions to take them over to the other side again. Well yes, but such an occurrence would be so improbable that it can be regarded as impossible. The arrow of time is also connected with what is probable.

In examining the directionality of the flow of time two explanations have emerged for the observed asymmetry of time. The first is connected with the irreversibility of actions and the second with probabilities. Both of these threads of argument need careful examination. Firstly, to look at the irreversibility of physical processes, sand falling under the influence of gravity is irreversible because there is no counter force to make the particles return to the top. The only way in which the sand can get back to its starting place is for someone to expend energy by turning the hour glass over again. Another irreversible process is the burning of a fuel like coal. Once combustion has taken place the particles that once were assembled in the black solid have been dispersed about the world in the form of smoke, dust, ash, and, in terms of the energy that bound the solid together, dispersed as heat. In fact, there is not a spontaneous change that occurs in the world that is reversible. A process can only be reversed by using energy, doing work, and dissipating heat.

The physicists of the late eighteenth and nineteenth centuries who examined the properties of heat, energy and work labelled this field of science thermodynamics – the dynamical properties of heat. In studying thermodynamics, processes could be thought of that were perfectly reversible and permissable according to the law of nature that states that energy can neither be created nor destroyed (First Law of Thermodynamics). Yet these processes are never actually found in nature. For example, a sample of gas could be heated by being placed in contact with a hot reservoir which would cause the gas to expand. The gas could then be removed and transferred to a position where it was in contact with a cool reservoir, where, if conditions were right, it would transfer that heat to the cold reservoir, and contract back to its original volume. Such a process entails only a transfer of heat from one place to another with the gas in the same state at the end as it was at the start. The reverse of this process would involve trying to expand the gas in contact with the cool reservoir, thereby extracting some heat from the reservoir and lowering its temperature slightly, then transferring the gas back to the hotter reservoir and compressing the gas so that it gave up the heat it carried back into this hot store. Once again, only a transfer has taken place. However, this reverse process never occurs and can never occur! If it did ships could power themselves by extracting heat from the oceans, use it to do work and expel the waste into the air. Free travel never sounded so good!

The answer to why heat cannot flow from a cold body to a hot one is that there is another quantity involved in physical processes which, it turns out, always increases if the process is spontaneous and which introduces an asymmetry, an irreversibility, into the physical situation. This quantity is called entropy and it is a measure of the disorder of the system under scrutiny. In spontaneous processes entropy, disorder, always increases, and if the more ordered, less chaotic state is to be regained, then some extra work is involved in reordering the relevant parts.

As a simple example, consider a new pack of playing cards, fresh from its wrapper. The cards are neatly ordered from ace to king in the four suits, but as soon as they are shuffled, or dropped on the floor, the order is reduced; the cards become mixed and to reorder them takes energy and time. In the case of gas in the hot and cold reservoirs, in the process of transferring some heat from the hot body to a cold body, calculations reveal an increase in entropy. The final state is more disordered than the initial state, so the two ends of the process are distinct and an asymmetry in time can be detected. In addition, the expansion and contraction of the transfer gas are spontaneous processes in the hot to cold direction, but in the reverse process the re-expansion of the gas could only be carried out by means of an external source of energy, as would the re-contraction, which indicates that the asymmetry is real and that entropy is a physical property.

The Second Law of Thermodynamics states that entropy increases in spontaneous processes (not all processes are spontaneous or irreversible). Examples can always be found which appear to disobey this law but, on careful examination, actually prove the rule. In a local situation entropy may well decrease. One example of this might be in cooling the interior of a refrigerator, where entropy certainly decreases, but only because energy is being expended. Electricity has to be generated by burning fuel, and the smoke and ash are certainly more chaotic than the original fuel was, so that the total entropy of a larger system than the refrigerator alone will have increased more than the local decrease in cooling the refrigerator interior. Ultimately, the total entropy of the universe has to be taken into account.

To again return to the gas clock, it becomes clear that when the aperture between the two chambers is opened the natural diffusion of the gas into the whole volume is a spontaneous process and that the entropy of the final state of the gas is greater than the initial state: the molecules are in a more disordered condition. However, entropy is not the only reason why the reverse process does not occur and the improbability of the situation has also to be considered. It seems highly unlikely that all the molecules will, of their own accord, go back to one chamber. Such a probabilistic argument can be worked out in detail statistically and a reliable probablility assigned to the chances of this unique event occurring. This sort of argument implies that although this event may be highly improbable, it nevertheless can occur, which seems rather different in approach to the ideas of thermodynamics and entropy. An example using a pack of cards helps put this into context rather more clearly.

In a pack there are fifty-two cards to which an order can be assigned (such as ace to king, hearts, clubs, diamonds, spades). What is the probability that given a proper shuffle the cards would end up in the order originally given? The case is a much simpler analogy than the molecules of gas in the clock, for here there are only fifty-two cards instead of something like 10^{23} molecules. The calculated probability turns out to be one chance in 10^{68} attempts. To put that into perspective, imagine every man, woman and child on this earth, each with a pack of cards, shuffling them once every second. How long would it take until someone shuffled a pack into the right order? The answer is about 10^{50} years. As the universe is only about 10^{11} years old, all those people would have to keep shuffling those cards for about 10^{39} lifetimes of the universe! To ascribe the word impossible to such a situation is, in the light of that answer, not a misuse of words.

Despite the absurd odds that may originate from a probabilistic treatment of a physical process like that described in the gas clock there nevertheless turns out to be a remarkable connection between such a treatment and the concept of entropy. In calculating the probability for the whereabouts of each and every gas molecule, the mathematical expression is derived from the same form as that for entropy. By adjusting the scaling factor the two calculations can be made identical. This scaling factor turns out to be one of the universal constants (Boltzmann's Constant), which suggests that the identity between the two expressions is significant. Entropy increases for those processes that are probable, and time flows in the direction of increasing entropy.

Sir Arthur Eddington first suggested that the increasing entropy in the universe could be the cause for the observation of time's arrow. The reason is because a later moment can always be distinguished from an earlier one by testing for the change in entropy. It even seems possible, at first glance, that entropy could be intimately connected with time, as if particles of time became 'used up' and hence converted into units of entropy, increasing as the history of the universe passes along. Indeed this idea seems even more attractive when such a conversion process is considered: the actual process of flow might consist of chronons turning into entropy units. But, alas, entropy does not increase in every process, and indeed can decrease locally. Perhaps such instances would account for timelessness and even for time reversal, but it is hard to think of time running backwards inside a refrigerator! The theory would not easily explain the connection of entropy with probability either. When the improbable did happen, would time flow backwards? No, the only correspondence between time and entropy that has any plausibility is in marking the arrow of time in one direction and there are limitations to that picture. It is satisfactory on a large scale but the lack of uniformity of entropy increase makes it difficult to connect with either the flow of time or the apparent uniformity of its one way flow.

My final comment on Eddington's suggestion is to remark on the consequences he foretold of the universal increase in entropy. The examples of both the gas clock and the transference of heat have shown that the end result of a spontaneous process is to edge things toward uniformity. In one case the density of gas molecules is more homogeneous over a wider volume of space and in the other case the hot and cold reservoirs are a little closer together in temperature difference. As the universe proceeds, and entropy increases, so too will disorder increase and the differences between things decrease. A hot star, embedded in cold empty space is a highly ordered object, while empty space ideally is homogeneous, isothermal and uniform. However, it is toward this uniformity that the universe is slowly tending. Such an unspectacular ending to universal evolution is called the heat death of the universe, because everything will end up the same temperature. Entropy will have won out and time will cease; there will be no longer be any way to distinguish between before and after.

It might appear that if the whole universe became homogeneous, isothermal and totally uniform, then it would be in a more ordered state than when it was full of contrasts, but this is very much a matter of definition within the subject area, jargon if you like. The increase in entropy really means the unavailability of energy to do useful work because it has been downgraded. Work can only be done when there are differences, in temperature, gravitational field strength and so on; entropy is a measure of the downgrading of these differences and is more disordered in that special sense. Alternatively, it can be seen as disordered like the shuffled cards, because the energy that went into their ordering has now been wasted and the objects can only be re-ordered by the expenditure of yet more energy.

The concept of entropy has not included anything about the flow of time, although it provides one view of why time might flow in one direction. To resist the unremitting universal increase in disorder work must be done, energy expended, in order to maintain an ordered state at the expense of creating more disorder somewhere else. That is how all systems, including living things, seem to 'disobey' the Second Law of Thermodynamics, but even then time takes it toll and the battle is won eventually by entropy.

Most biological processes are irreversible and it is this fact that has often been used to maintain that man's perception of time's flow is inate, a part of his biological make-up. A geological equivalent of this mechanism would be the deposition of sedimentary rocks, whose constitution and arrangement forms a 'memory' of some physical changes that occurred with time, during the history of the earth. These changes, too, were irreversible, and tended towards the maximum probability. The rocks were formed by entropy increasing processes, and their subsequent erosion by the weather continues this inexorable march toward the heat death of the universe.

Thermodynamics is concerned with the gross behaviour of matter and energy and is therefore probabilistic. It has nothing special to say about the individual atoms or molecules that partake in the gross behaviour. But the field of sub-atomic or quantum physics is another branch of physics where the use of probabilities is also supreme. Curiously enough, when scientists investigate the smallest chemical particles, atoms, which in turn are composed of even smaller and more 'fundamental' particles, then statistics are also used to describe their behaviour. The reason why this is so can only be answered by briefly discussing the structure of atoms.

In the early years of this century scientists such as Neils Bohr and Ernest Rutherford were investigating the properties and structure of atoms. Rutherford had shown that atoms contain a central, compact nucleus and Bohr went on to propose that electrons circled this nucleus rather like planets around the sun. Now according to the classical laws of physics, charged particles like electrons could only travel in curved paths at the expense of radiating away energy in the form of light. To encircle a nucleus meant that an electron would be constantly accelerated and would have to lose energy. The consequence of this would be that it would spiral in toward the atomic nucleus. Indeed, calculations indicated that this collapse would occur in a small fraction of a second.

To solve this problem, Bohr turned to an idea originally proposed, under different circumstances, by the German scientist Max Planck. What Planck had been faced with was a breakdown in the applicability of the wave description of light; his solution to the problem revolutionized physics. The difficulty concerned the energy associated with radiation from a hot body, like a glowing piece of coal, a furnace interior or even a star. The energy carried by wave motions is associated with the amplitude of the wave, the height of the wave crest or the intensity of the light. Light given out by a hot body could be thought to contain the wavelengths that could be fitted into the space of the body. A box, like that shown in Figure 4.1, which could be a furnace interior, is heated to a certain

Figure 4.1   *Fitting wavelengths into a box. If only whole wavelengths are permitted there is an upper limit to the size of wave permissible.*

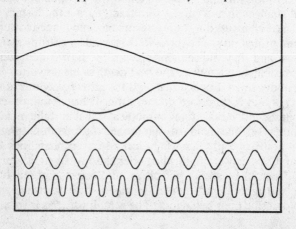

temperature. A wavelength the same size as the box can be fitted into it and so can as many shorter wavelength waves as liked, an infinite number of them in fact. As shorter wavelengths contain more energy an infinite number of short waves in the box would fill it with an infinite energy. This problem of description was called the ultraviolet catastrophe and it contains two aspects of the problem of description of waves.

Waves can be described by their wavelength or by their frequency. These two parameters are inversely related, so if the wavelength is zero the frequency becomes infinite, or if the frequency becomes zero the wavelength becomes infinite. Part of the problem facing Planck is related to fitting wavelengths into the box, although fitting frequencies would also lead to the same problem. But the difficulty of the frequency/wavelength relationship is that it imposes a limitation on describing some properties of waves. For example, in drawing a scale illustrating the range of radiation in the full electromagnetic spectrum should one use frequency or wavelength for the scale? In either case one end of the scale must go to infinity, unless one representation is changed to the other somewhere midway. To return to Planck's problem, he knew, from experiments performed, that at short frequencies there was a cut off in the radiation from hot bodies. He therefore proposed that energy was related not to the intensity of the light but to its frequency. As the frequency increased, more energy had to be put into it, and if there was only a certain amount of energy to use, corresponding to a particular temperature, the higher frequencies could not be powered. The idea is really not that surprising because frequency tells how many waves have to be packed into a unit time. The more waves packed in requires more energy, so as the wavelength becomes shorter those extra waves have to be packed into an ever decreasing space. Planck's solution to the ultraviolet catastrophe was simple but went against the classical description of phenomena. Furthermore, it meant that energy divided by frequency came in basic packets, called quanta, and that each quantum had the dimensions of energy times time. This unit is called action.

The remarkable thing about action is that although it has dimensions it comes in units and so can be counted. One can have three packets of action, or two, but not two and a half. Action contains the dimensions of mass, length and time and yet is not expressed in units such as kilograms, metres or seconds. Action in physics is really like action in life, you *do* something, like walking, and the action is a single unit, even if it is made up of parts. The quantum, which underlies modern physics, is not, as is so often expressed, a packet of energy, but a unit of action, which contains energy and time. In pursuing the description of the quantum theory yet another aspect of time is encountered.

The idea of the quantum had already led Einstein to suggest that light, too, could be quantized; that is, it comes in particles, which he called photons. Now it was Bohr's turn to make use of the quantum in describing a stable configuration of orbiting electrons around the nucleus of an atom. The proviso was that the electrons could only exist at particular, fixed orbits, that is, the energy structure of the atom was not continuous but made of discrete steps. In a stable, permitted orbit, an electron could remain circling the nucleus indefinitely without radiating energy, and could go to higher or lower energy orbits by gaining or losing the right amount of energy, probably in the form of light. Such a description thereby enabled Bohr to explain the characteristic features of the spectrum of light emitted by an atom.

In trying to understand the characteristics of electrons and their configurations in atoms, which led to a clear understanding of the nature of the chemical elements, Wolfgang Pauli proposed that no two electrons within an atom could possess identical energy states. Even where two electrons seemed to share an orbit he showed that in fact the electrons would be spinning in opposite directions, hence smearing out the sharply defined energy level within the atom.

At this time the concept of the wave/particle duality of light was accepted and also applied to small particles, which de Broglie suggested could equally well be thought of as waves. Indeed, in the late 1920s Davison and Germer demonstrated that what are thought of as particles can also behave like waves with the diffraction of a beam of electrons. In terms of understanding the atom, this wave/particle duality led to a wave formulation for electrons within the permitted orbits. The hardness and precision with which particles were pictured gave way to the rather nebulous idea of a wave. The combination of waves and particles is often depicted by use of a wave packet, as shown in Figure 4.2, where the wave is confined to a small region and so combines its waviness with particle-like properties as well. The particle is not totally localized and yet the wave is not limited in space/time.

It is not so easy to say exactly where the electron is inside an atom if it is wavelike and so a statistical or probabilistic method was developed to indicate that the electron is more likely to be here than there. Such an approach to the nature of electrons and the structure

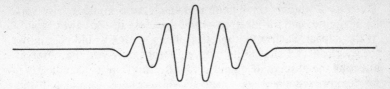

Figure 4.2   *A wave packet.*

of the atom led to rapidly increasing understanding and mathematical descriptions of sub-atomic physics, with electrons becoming
described as 'probabilistic clouds of smeared out electric charge'.
Particles were portrayed by wave functions, and, as quantum
physics became established, the key equation was Schrödinger's
wave equation which dealt with the probability distribution of the
whereabouts and motions of the electrons and other fundamental
particles. The simple scheme established by Neils Bohr of a solar
system kind of atom became, in the space of twenty years, a highly
complex, three dimensional, series of interpenetrating spaces where
electrons were more likely to be found.

Indeed the development of physics has been described as the
progression from waves of matter being transmitted by a medium (a
wave on a string) to waves of something being transmitted by
nothing (light) to waves of nothing travelling in nothing (electrons in
atoms)!

Perhaps one of the most important results that emerged from the
new quantum theory, which not only changed the face of physics but
also unravelled some of the mysteries of the sub-atomic micro-
universe, was that due to Werner Heisenberg. At the level of the
smallest conceivable particles, energy packets in the form of
photons of light can have a physical influence on the behaviour and
motion of such particles as electrons. Heisenberg realized that to
measure the position and motion of, say, an electron meant
scattering at least one photon off the particle in order to see it (we
see by photons entering the eye). However, the wavelength of
visible light is orders of magnitude larger than the size of an electron
and just as radio waves sweep us by without us affecting them or
they us, so visible light would fail to interact with an electron. An
electron could not be *seen*. To overcome this difficulty Heisenberg
reasoned that radiation whose wavelength is about the same size as
an electron must be used, but small wavelength means high
frequency and, as Planck had shown, therefore high energy.

Throwing a suitably small photon at an electron would involve very high energies and the electron would therefore be shot away rapidly by the impact. Such a photon/electron interaction would both move the electron and change its motion, so not only was the observer affecting the observable quantity but affecting it in such a way that it no longer remained as observed. Heisenberg formulated this awkward but fundamental limit to observational capability as a physical principle which stated mathematically that whenever two interdependent quantities, such as position and motion, of a particle are to be measured, there will be an uncertainty in each measurement and the product of the uncertainties will always be larger than a particular, given value.

What Heisenberg's Uncertainty Principle is saying is that precision in one measurement can be gained only at the expense of losing information about the other quantity. The more precisely the position of a particle is determined, less is known about its velocity. In fact, the number that the uncertainties will always be greater than is very small (10^{-34} Joule seconds), but nevertheless quantum theory contains detailed limits to knowledge. There can be no absolute certainty at the sub-atomic level. The theory is essentially indeterminate.

In addition to the pair of quantities, position and velocity, the Uncertainty Principle operates for the two properties of energy and time. Energy/time is action, which lies at the basis of the quantum theory. Furthermore, the number that uncertainties always exceed turns out to be Planck's Constant, 10^{-34} Joule seconds, which is the size of one action unit. And now, action involves time in uncertainty. If a scientist wants to gain knowledge of the energy of a sub-atomic force, for example, he must lose out on the accuracy with which he can determine the time interval over which that force operates. It is this example of Heisenberg's principle that accounts for some strange and yet fundamental properties of this description of time associated with the matrix that seems to lie at the basis of the physical world. As with the macroscopic universe, the microscopic level of reality also offers unusual perspectives on time.

Before exploring the temporal aspects of quantum behaviour, I must make a few remarks about what are called the fundamental particles. At the beginning of the twentieth century scientists were just realizing that atoms had some inner structure. Until then atoms were regarded as being the 'ultimate' particles from which all things were made, and that each chemical element could be reduced to simple atoms. However the discovery of the electron and then the proton showed that atoms themselves could be further reduced to these more fundamental particles, and that indeed the distinct atoms of each chemical element were distinct because of the different number of electrons and protons that went into their composition.

Unfortunately, sub-atomic physics turned out not to be as simple as it first looked, and more and more elementary particles were postulated to account for increasingly detailed and complex properties of atoms that could not be explained by protons and electrons alone. The particles that theoreticians required to balance the properties of atomic structure and to account for the observed behaviour of the known particles led experimenters to search for and thereby find a whole range of new 'fundamental particles'. The particles predicted were found, and some unexpected ones also discovered, which led to further expectations of still more new particles. There are over 200 known fundamental particles, more than double the number of known distinct chemical atoms, which rather discounts their description as fundamental. Most of them only exist for minute fractions of seconds, far too short to measure, and are in no sense 'stable'. I shall say a little more about those shortly.

In the late 1920s the Cambridge physicist Paul Dirac realized that the equations describing the sub-atomic particles were symmetrical with respect to several of their basic properties, such as electric charge. He proposed that for every particle which existed there was an anti-particle which was its mirror image. For example, the anti-particle for the electron, later discovered and named the positron, was identical to the electron except that it carried a positive rather than a negative charge. In this way the number of fundamental particles doubled instantly and most of the anti-particles have been discovered. For those that have not, such as for the neutron and the photon of light, it is said that the particles and the anti-particles are identical.

To review, the quantum theory originated by discovering action, and was developed by the study of the sub-atomic world. Because the basic interaction of light with sub-atomic particles does not permit measurements with complete certainty to be made, a description of fundamental particles and processes is restricted to statistical methods. Hence quantum theory is a probabilistic description of nature, not as in the macroscopic world because of the sheer numbers of particles, but because of an inherent property of indeterminism in describing physical reality. This final restriction on the attainability of absolute knowledge has worried many scientists, but the current consensus is that the quantum theory is the fundamental theory of nature. There are other opinions, such as

David Bohm's, which argue that the phenomena of particles and so on are but the explicit manifestations of a more fundamental order, an implicate order, of reality.

The final ingredient of this story involves the forces acting between particles at the sub-atomic level, and, after looking briefly at the current description of the way these forces operate, all these aspects of quantum physics can be brought together to obtain a sub-atomic perspective on time.

Both the gravitational force, which operates between masses, and the electromagnetic force, which operates between electric charges or moving currents or magnets, are familiar everyday experiences. However, these two forces play no practical role in the sub-atomic world where two less familiar forces take over. One of these is called the weak force which, for example, is involved in radioactive decay, and the other, amazingly, is called the strong force which acts between particles in the atomic nucleus. There has always been a problem in science over forces because they act 'at a distance'. The gravitational pull of the sun on the earth acts just like someone pulling on the string while swirling a stone round their heads on the end of the string – except that for the sun/earth system there is no string. The force acts through empty space and this has always been most unsatisfactory to scientists, who like to know by what mechanism something works. By considering the workings of the nuclear forces scientists replaced the doubtful 'action at a distance' explanation with a much better understood and acceptable answer.

Consider the case where two particles, say, electrons, approach each other, and, sensing their mutually repulsive negative electric charges, repel each other and separate. We say a force has operated to change their respective motions. As the particles never touched each other that force acted 'at a distance'. Another way of describing it, however, would be to say that one electron 'throws' a particle at the other electron, which is 'caught' by that one and this exchange of energy, via the intermediate particle, alters their paths. The system would operate rather like ice skaters tossing a heavy ball to each other. When one skater throws the ball he recoils from the throw and changes direction on the ice, just as does the skater who catches the ball. The situation is described in Figure 4.3, where the diagram on the left shows the repulsion behaviour of two particles, looking like action at a distance, whereas the picture on the right includes the exchange particle which executes the repulsion.

Now look at the illustrations again. What sort of diagrams are they? Space/time diagrams. Time is plotted up the vertical axis and

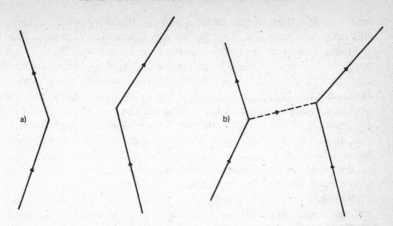

Figure 4.3 *The repulsion of two particles. In case (a) the repulsion looks like 'action at a distance' while in case (b) it is due to an exchange particle.*

space along the horizontal. In particle physics space/time diagrams come into play yet again. Because the Nobel prize winning physicist Richard Feynman first used space/time diagrams to describe particle interactions, and related the diagrams to the mathematical expression describing the interaction, such pictures are called Feynman diagrams.

To consider Figure 4.3 as a particular example, the particles might be electrons and the exchange particle would then be a photon of light. The force then being described is the electromagnetic force of repulsion between two charged particles. Instead of each experiencing the effect of the other's negative charge, what has happened is that one has emitted and the other absorbed a photon, and the mutual exchange of photons then constitutes the repulsion. (Although only one exchange is shown in the diagram a series of cross exchanges between the electrons occurs.) This mechanism can account for the operation of all forces, and experimental particle physicists study the interactions of fundamental particles partly in order to examine exchange particles. But from where do these exchange particles come and how do particles know when to throw them out?

To answer the first part of this question consider the mutual repulsion of two protons, which is executed by the exchange of a particle called a pi-meson or pion (symbol π). The family of particles collectively called mesons are in fact all exchange particles for the strong nuclear force, just as photons act as exchange particles for the electromagnetic force. The pion has a mass about one seventh that of the proton and so throwing such a heavy particles accounts for such a strong interaction between two protons. The proton-proton repulsion in Figure 4.4 shows what happens. But surely doesn't one proton end up one seventh lighter, and the other one a seventh heavier? Yes, but only momentarily because each proton is mutually exchanging pions. How can the proton conjure up a pion from nowhere? The answer to this lies in Heisenberg's Uncertainty Principle, where, it will be remembered, the product of the uncertainty in energy multiplied by the uncertainty in time must be greater than a particular small number. If the time element in this relationship is small enough, the energy part of it can be made as large as necessary and Einstein showed us, as a consequence of the Special Theory of Relativity, that energy and mass are equivalent. Hence, if enough energy is available it can be converted into mass, and thus create an exchange particle provided it is done quickly enough. In other words, if an observer reduces the uncertainty in time in order to measure the energy of a sub-atomic system, the uncertainty in the energy increases so much that he could be seeing one particle or that particle plus another. Once again, the precision of measurements is in question. The analogy often presented for this process is of the shop assistant who 'borrows' from the till. Providing he returns the money very quickly he won't be caught out, and the more he takes the more obvious it will be, so the time interval will be shorter during which the borrowing is safe. (It is a curiously immoral analogy and that in itself is interesting to a description of processes that are intended to be

Figure 4.4　*Proton-proton repulsion by the exchange of a pion.*

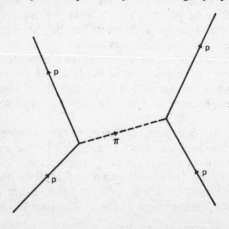

very fundamental and yet actually do sound like cheating.) It is the same with particles. A proton can produce a pion out of the uncertainty in its energy/time budget provided it does so for a very short time only. Indeed when time intervals become short enough almost anything can happen; matter can be created, forces occur and so on.

The second part of the question involves the protons knowing when to exchange particles. Particles created in the short time intervals allowed for by Heisenberg's Uncertainty Principle are often called virtual particles because they only exist within the confines of that small time interval. The more massive the virtual particle the shorter the time interval and the closer the emitting particles must be. This explains why the strong nuclear force operates only over very short distances, because the exchange particles are massive, whereas the electromagnetic force operates over comparatively long distances indeed, because photons are massless and so require very small amounts of energy to be created. It looks as if, then, the operation of the physical world depends on very small time intervals.

Although particle interactions or forces are due to the exchange of virtual particles, there is nothing in quantum theory to prevent virtual particles from being created at any time, providing they do not exist for long. A proton might be found on its own emitting a pion and then reabsorbing it, as shown in Figure 4.5, and indeed any such virtual particle creation that can be imagined can in practice

Figure 4.5 *The creation of a virtual pion by a proton.*

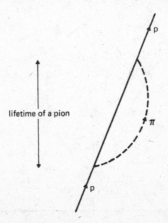

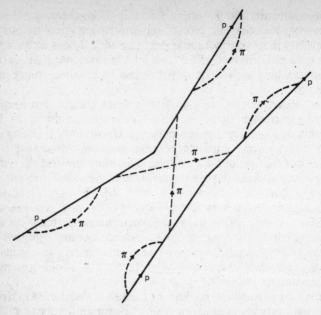

Figure 4.6 *The repulsion of two protons by pion exchange.*

Figure 4.7 *A vacuum diagram. This illustrates the creation and their subsequent annihilation from the void of a proton (ρ), an anti-proton (ρ̄) and a pion (π).*

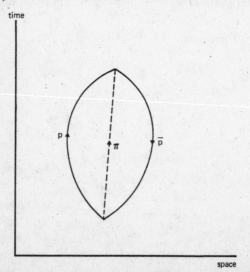

occur, and does occur, within the rules of quantum probability. Instead of imagining particles like protons sedately moving along, a more realistic picture would be that they are surrounded by a haze of seething virtual particles, none of which exist for more than vanishingly small times. Such a particle, when approached by another, does not need to 'know' a repulsion is required; the cloud of virtual particles will overlap and particle exchange will occur, as illustrated in Figure 4.6.

One final effect of virtual particle production to be considered is, of course, the logical extension of the creation of particles out of particles, namely the creation of particles out of nothing. If there is an empty void between particles then out of that void, or vacuum, emerge massive particles, but of course only for a short time, before they disappear back into the vacuum. The whole point of virtual processes is that they occur so quickly that scientists can never catch nature doing them. Whenever the manager looks in the till the money's still there. Such a vacuum process is shown in Figure 4.7. In this illustration a proton (ρ), a pion (π) and an antiproton ($\bar{p}$) are created out of nothing and rapidly reconverge on each other to mutually annihilate. Such ethereal activity has led physicists like David Bohm to talk about the enormous energy content of what we call the vacuum, and Fritjof Capra to compare our descriptions of such events with the words of Chang Tsai: 'When one knows that the Great Void is full of ch'i one realizes that there is no such thing as nothingness.'

The particle physicists have discovered over two hundred fundamental particles, the vast majority of which are virtual particles, with incredibly small lifetimes, around 10^{-22} seconds. Again, to use an extreme case (as was done with high speeds in relativity) time displays strange characteristics. As a limit to the energy associated with virtual particles is approached so a limit to the length of meaningful time intervals is approached. That limit seems to be about 10^{-24} seconds, roughly the time it takes light to cross from one side of a proton to the other. Is this the chronon? Can events take place in less time? Or is this a meaningless question because scientists can never see beyond such a scale? It is interesting to speculate whether the questions have to do with explaining time, or rather with something that is labelled time in order to explain events. Does time at this quantum level exhibit precise properties or display definite behaviour?

To return to Figure 4.7, the arrow indicating the motion of the anti-proton in that diagram was downward, in the opposite direction to that of the proton. But this is a space/time diagram and time is always supposed to flow forward, so why does the arrow point backward in time? The answer comes from the mathematical description of anti-particles. Anti-particles are mirror images of 'normal' particles with one characteristic of the opposite sign, such as the electron (e⁻) and the positron (e⁺). However, they can be equally well described as normal particles moving backward in time. A positron is either a positively charged electron or else it is an electron moving backward in time. If the description allows this to be a property of the behaviour of particles then that is how they may well behave. It is impossible to tell the two cases apart: they are equivalent.

Figure 4.8 shows the annihilation of an electron and a positron, with the emission of a photon (a gamma ray, γ). Equally well an electron could be reversing its direction in time, emitting a photon in so doing. The evidence of such an encounter can be seen by looking at the photograph of the particle tracks in a bubble chamber. There, the annihilation appears much as it is shown in the diagram – it is impossible to tell whether the positron was a positron or a time-reversed electron. This sub-atomic time reversal means, of course, that the order of events cannot be determined. Figure 4.9 does not indicate whether a photon created a positron/electron pair, which lead to the positron annihilating with a second electron to create another photon, or whether one electron passed through a time reversed period, with the emission and absorption of photons in the process. As the two versions of the story have different causal

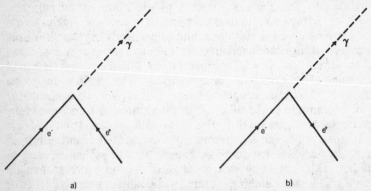

a)   b)

Figure 4.8   *The annihilation of an electron (e⁻) and a positron (e⁺) with the emission of a gamma ray (γ) seen either as time forward particle collision (a) or an electron reversing its direction in time (b).*

and sequential scenarios, then what is regarded as causality cannot hold at the quantum level of reality.

This chapter began by asking questions about the flow of time and the unidirectional nature of that flow. At the sub-atomic level, it seems not only that time may flow forward and backward but that, because of the latter behaviour, the idea of flow itself becomes confused. For example, the case shown in Figure 4.9 is difficult to examine from the point of view of flow because time direction is being switched about. Flow implies strict causality. Causal connection is an uncertain principle in quantum physics and the benefit of use of the concept is doubtful.

However, a different question comes to mind about the arrow of time from the examination of anti-particle time reversal: namely why does time's arrow flow in one direction at the macroscopic level if at the microscopic it reverses? The answer to that question has been sought by cosmologists in their attempt to find out why the universe is apparently dominated by matter rather than equally shared out between matter and anti-matter. Their present solution to the problem lies in the particular particle interactions that occurred in the very early stages of the universe which, they tell us, led to this asymmetry. The answer to the question lies, at least in part, in the observation that the universe consists primarily of

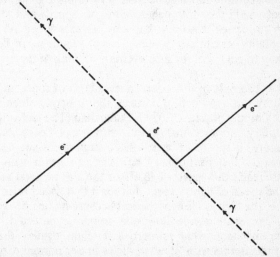

Figure 4.9   *A complex interaction between electrons (e^-), a positron (e^+) and photons of gamma radiation (γ).*

normal particles with only a minority of anti-particles. This asymmetry gives another reason for time's arrow.

But to return to fundamental particles, reconsider Figure 4.7. There are three virtual particles created out of the vacuum at position one, which mutually annihilate and return to the void at position two later in time. But if the anti-proton is considered as a proton moving backward in time, there is another version of the story. A proton and a pion are created at 1 and move towards self-destruction at 2. At the instant of their annihilation a proton is created which moves backward in time to annihilate at 1, the instant the proton-pion pair are created. Or the proton is switching from 1 to 2 to 1, with the creation and destruction of a pion at each change of temporal direction. Nevertheless, whichever way this diagram is read positions 1 and 2 are inextricably mixed up. The events at 1 could not occur if they did not 'know' in some way of the events at 2, occurring later in time! This interpenetration of space and time, which is a feature of such a description of reality which is commonly found in quantum theory, makes for a feeling of unity throughout space and time. Every place is interconnected with all other places and all other times by these virtual processes whose end result is somehow built into the initial event. Such a viewpoint makes it very difficult to see time flowing, because flow implies going from here to there, and the interpenetration of space and time by quantum processes dissolves that flow. Events 'happen' in some sort of totality, essentially independent of sequence, of flow. The easiest way to get a feel for such events is to think of them as simply *being*.

Another example of this interconnectedness of space and time is provided by the hypothetical particle, the tachyon, which was first postulated by the physicist E.C.G. Sudarshan. The tachyon has the extraordinary property that it can *only* travel faster than light. Furthermore, it speeds up as it loses energy and slows down as it gains energy. Just as ordinary particles can never gain enough energy to reach the speed of light, so too the tachyon can never obtain the speed of light, except that for it that speed is its slowest. Of course, the tachyon is hypothetical because no one has ever seen one, or can even decide how one could be detected. Particle physicists speculate what properties it might display that could enable it to interact with 'slow' particles in our normal experience. But despite the ethereal nature of these objects, speculation concerning the tachyon is enlightening.

Imagine that at the origin of the universe, in the Big Bang explosion that initiated it, tachyons were created. These particles would have travelled out through space faster than light and became faster as they lost energy due to the expansion of the universe (tachyons speed up as they lose energy, just as normal particles slow down; that is, they get further away from the speed of light). There comes a time, however, when they are travelling so fast, essentially infinitely, that they reach a minimum energy state, just like being at rest for a normal particle. They cannot get any faster at this time and yet the universe continues to expand, so all they can do is move backward in time until they end up back at the origin of the universe, but a long way away! This extraordinary story is pictured in Figure 4.10.

The curve of the tachyon path on this space/time diagram is always more than 45° from the time axis, because of its faster than light property. It reaches the time of minimum energy at t_{min} and then it has travelled a distance d from its origin in the big bang. By the time it has returned back in time to that origin it has gone twice that distance (2d) and so connects vast astronomical distances at the 'same time'. Another way of interpreting this diagram, however, would be to think that at the origin of the universe a tachyon and an

Figure 4.10   *The space/time diagram for the faster-than-light particle, the tachyon. At time t_{min} the tachyon has 'slowed down' to an infinite speed and can only continue backwards in time, connecting position of its origin with its destination at the same time.*

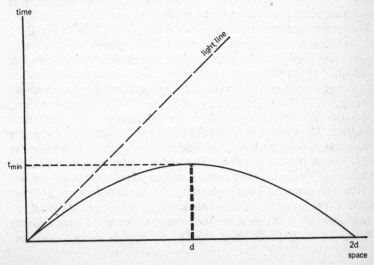

anti-tachyon were created at the same moment but separated by a distance 2d. As the particles travelled out across the universe, each was speeding up and the two collided at time t_{min} and place d resulting in their annihilation, which would be a gentle affair as each particle would be in its lowest energy state. Such a scenario once again correlates different times and places in such a way that the barriers we encounter in both space and time seem to no longer exist. Particles interpenetrate vast distances, connecting the very limits of the universe with each other in this strange way.

In describing this account of the tachyon I said it reached a point where it could only continue by going backward in time, and by now this concept does not seem so bizarre. To examine this idea a little more closely, the tachyon is always travelling faster than light and yet is always moving forward in time. What would a tachyon observe (if it could observe anything) in the way of a light signal? Imagine that a television programme was being transmitted across interstellar space. The tachyon would overtake the light wave and hence see the end of the film, the middle and then the start. To all intents and purposes it would 'see' things reversed in time. It would see trees spring up from decayed matter, collect up fruit from the ground and transform them into flowers that would, like its leaves, reduce into the branches as the whole tree shrunk down into a seed. And yet the tachyon would be moving forward in time! In fact, the anti-particle, the anti-tachyon, would be the particle that viewed things in a forward sequence, as it moved backward in time but still faster than the speed of light. Perhaps this is not so surprising as tachyons seem like anti-particles in the mirror of the speed of light; their postulated existence questions the notion of time reversal.

One final point concerns time travel. The explorer in H.G. Wells' novel *The Time Machine* built a device that moved through time but remained in the same place; no known temporal phenomena can do that. In all cases of time reversal in the quantum domain, position changes as well as point in time. Of course all objects travel in time, because they never stand still on the path through the space/time diagram. Time travel really means transcending that limited sense of temporal journeying, either moving forward faster in time, like the tachyon, or moving backward in time (which may well be like the tachyon as well). In either case, an object would be transported out of the light cone into the region that is neither here and now nor past and future. Quantum theory can allow this to happen, but only to 'fundamental' particles. Even the form of time travel I have referred to in relation to black holes strictly only applies to such particles.

The effect relies on quantum theory with its uncertainty overcoming improbability. Time travel in a practical sense is still elusive and time still seems trapped within the barrier of light.

This chapter has considered the flow of time at a macroscopic level, which uses a statistical or probabilistic approach in order to be able to describe phenomena. Such an approach leads to a discovery that the flow of time points to states of maximum probability and hence to a one-way flow. However, the property of entropy is not really an indicator of flow, so much as a direction indicator. At the macroscopic level, then, one can learn much about uni-directional flow but not much about flow as such.

At the microscopic or sub-microscopic end of things, the descriptive processes also have to be statistical and probabilistic, not because of the vastness of the system, but because of the uncertainty in precise knowledge available at this level of reality. But in this case, the statistical approach does not give any real indication of a one-way flow. At the quantum level flow abounds. The vacuum seethes with flowing particles, but the temporal direction is not confined to forward. Flow can go anywhere. Indeed the abundance of flow is so overwhelming it virtually ceases to exist at all. The two extremes presented in this chapter have, then, given some indications of what time forward and time reversed may mean, even if the meaning that emerges may be that such ideas are due to local confinement in our personal light cones, restricted by the speed of light that seems so bound up in the study of time.

5

TIME AND THE UNIVERSE

To look up at the stars in the heavens is to look back in time to see past history – providing it took place far enough away. This is a strange and intriguing thought that deserves a little careful attention.

Because light travels at a finite speed, the light reaching earth from a distant star has taken a certain time travelling through space. The photon entering someone's eye was emitted by the star in the past. The farther away the star the longer it has taken for the light to reach the earth; the star is seen as it was a certain length of time ago. That is, to see a distant star is to look back along the light cone. As the cone spreads out, the light comes from farther back in time and from an increasingly greater distance (Figure 5.1 reproduces the light cone from Chapter Three). This means that the sun is observed as it was eight minutes ago down the light cone. A nearby star may be seen as it was a few years ago and the Milky Way a few tens of thousands of years ago.

The speed of light and the distances to astronomical objects are indicators of how long the light from distant objects has taken to reach the earth. The farthest away the naked eye can see is to the Andromeda galaxy (also labelled M31) which is so far away that its light takes two and a half million years to reach the earth. Telescopes extend the eye and enable fainter and fainter objects to be seen as they were thousands of millions of years ago. What is seen may no longer even be there now. But what is meant by 'now'?

The light from a distant star has been travelling from that star along the 45° line of the space/time diagram when it interacts with the eye at the moment called *now* at the same time as light from more distant and closer-by objects also enters the eye. All the light that comes into the eye 'now' has a different history. It can be argued that nothing is seen now because everything lies at some distance and therefore is distant in time as well as space. The reader of this book does not read this page 'now', but in the very recent past. Now is when photons enter the eye, and those photons may have come recently from nearby or they may have travelled a long time from a long way away. Past history is, in this context, 'now'; or, more accurately, seeing is a puzzling mixture of past and present. Indeed, the difference in perception of this page and a distant galaxy is only one of degree and not of kind. Both things are seen back in time and yet both are seen now.

To confuse the situation still more, no time passes at all for light during its travel. It arrives at the same time as it left. From the photon's point of view it enters the eye at the same moment that it left the distant star, hundreds of years ago. From one point of view the observer sees back in time, but from the photon's vantage point it really is seeing now! And this all arises because of the interconnection of time with light.

The last chapter discussed the idea of virtual particles being emitted and reabsorbed during the passage through space and time by more 'stable' particles: the exchange of these particles constitutes the forces observed between particles. If there is no nearby particle to cause an exchange then the virtual particle is reabsorbed. The examples were primarily concerned with the nuclear forces but the electromagnetic force, in which light particles, photons, are exchanged, operates on the same principle – except over much larger distances and on a universal scale. The emission and absorption of

Figure 5.1 *A Minowski diagram showing the light cone connecting here and now with the past and future. Places and times outside the cone are causally unconnected with here and now.*

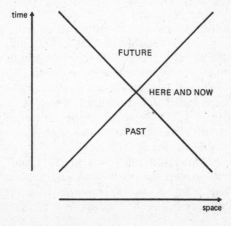

light, for example light emitted by a star and absorbed in the eye, can be considered a particle exchange. If no eye, or anything else for that matter, intervenes, the photon will travel on forever. But Einstein's General Theory of Relativity, which describes the structure of space and time, indicates that linear space is really curved and the photon, travelling in a 'straight line' will eventually end up back where it started, just as does a traveller going in a 'straight' line around the earth. In other words, the photon, if it is not exchanged, is effectively reabsorbed, just as other virtual particles are, by the atom which released it, even though its path is on the scale of the universe. Of course, the time interval for such a particle to exist would be extremely long, but to the photon its reabsorption would be at the same time as its emission.

At this point it may be helpful to return to the observation of stars and galaxies in order to understand more about the arrow of time. It was not until the end of the 1920s that the spiral galaxies were fully recognized as being 'island universes', vast collections of stars, gas and dust, as is the Milky Way. It was soon discovered that there were, in fact, countless millions of other galaxies, all containing hundreds and thousands of millions of stars, spread throughout a vast universe.

In 1928 the American astronomer Edwin Hubble, who had been making careful observations of many distant galaxies, discovered what turned out to be perhaps the most important astronomical find of the century. What he found was that the light from all distant galaxies was systematically of longer wavelength than comparable light sources close to the earth. This means that in a comparison of light from a galaxy with light from a laboratory source every component in the spectrum of the galactic radiation will be displaced towards the red end of the spectrum with respect to the laboratory source. This red-shift of light was not only found for all distant galaxies but also the size of this wavelength displacement increased the fainter, and therefore more distant, the galaxy.

Red-shifts were not uncommon observations in astronomy, and had already been useful tools for around thirty years when Hubble made this discovery. There was nothing intrinsically mysterious about them as such. The effect is due to a relative motion between a source and an observer for any wave phenomenon, and is called the Doppler effect. This effect can be noted every time a motor bike rushes down the street, the pitch of its engine-sound falling as it passes one by. When a source of waves is coming toward an observer the waves catch each other up, and, conversely, the waves lag further behind if the source is receding. If the waves are light waves then the approaching light is blue-shifted and the receding source is red-shifted (see Figure 5.2). Astronomers had used the techniques for observing Doppler shifts in studying the light from stars and thereby detecting stellar motions and rotations, but the remarkable thing about Hubble's observations was that all galaxies displayed a red-shift; all galaxies were receding from this one, and furthermore, the more distant galaxies were receding faster than the nearby ones. The regularity of the increased speed of recession with distance led Hubble to formulate his observational law that the velocity of recession of a galaxy was proportional to its distance. The scaling factor of this effect is known as Hubble's Constant (shown graphically in Figure 5.3).

The obvious interpretation of this observation is that the universe is expanding. Although it appears that the Milky Way is at the centre from which all other galaxies are receding that view is misleading. The universe, it turns out, is expanding uniformly and from any galaxy it would appear that all other galaxies are receding. It is the space between the galaxies that is expanding, not the galaxies themselves that are moving away from each other. A useful analogy is to think of a currant cake being cooked in the oven. At first all the currants are fairly close but as the cake cooks the dough between the currants expands as the rising agent gets to work and hence the currants all move away from each other. In the universe the galaxies are the currants and space itself is the dough. Whichever currant is chosen as a reference, all other currants are moving away from it, and indeed the further ones will be moving away faster because there is more expanding space (or dough) between them.

Hubble's observations, leading to his law relating the red-shift to the distance of a galaxy, enabled astronomers to put a real observational constraint on their models or descriptions of what the universe is really like. The value of Hubble's Constant at the present epoch is the numerical constraint in such descriptions or models. Hubble in fact made an incorrect assumption in obtaining a value of his constant and there is still controversy over its exact value. The constant itself, as well as indicating how fast the universe is expanding, is also the inverse of the age of the universe, so determining the value of Hubble's Constant tells the probable age of the universe. The most successful model of the universe and its evolution is called the hot Big Bang model, which explains the universal expansion as being the result of a compact and hot initial

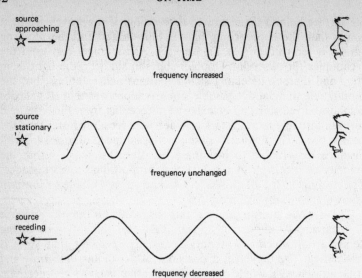

Figure 5.2 *The Doppler Effect. An observer sees the wavelength shortened, that is blueshifted, if the source is approaching. The observed and emitted wavelengths are identical if the source is stationary with respect to the observer, but the wavelength is increased, that is redshifted, if the source is receding from the observer.*

Figure 5.3 *Hubble's Constant, showing the linear relationship between the velocity of recession of distant galaxies, as expressed by the redshift, plotted against the distance. The circled dots represent the redshifts and distances of individual galaxies and the line is the theoretical connexion between them.*

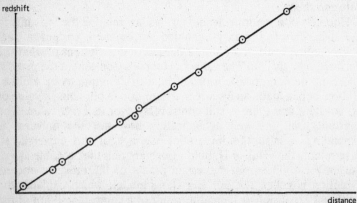

'cosmic egg' exploding at the beginning and the result of which is still seen in the recession of the galaxies. I am not going to argue about modern cosmological theories, but will instead use this simplified statement about the main current view of the nature of the universe to explain its consequence in relation to time. In practice, as far as time is concerned, the different cosmologies that have been suggested by theoreticians are less interesting than the agreed-upon fact that the universe is expanding.

Figure 5.4 shows the world lines of several galaxies, all of which

Figure 5.4 *The world lines of receding galaxies will be separating from each other. The line that cuts each world line at right angles links moments of 'now' that could be regarded as times of equivalent epoch in the history of the universe. Before NOW the lines were closer together suggesting the compactness of the universe at the possible origin.*

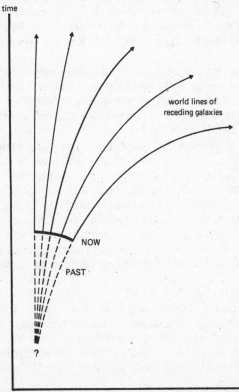

time

world lines of
receding galaxies

NOW

PAST

?

space

are becoming more separated in space as time increases, which is merely an illustration that the universe is expanding. Such regular behaviour of the universe and the objects comprising it can lead to the suggestion that epochs in the history of the universe can be dated and thus a system of time not unlike Newton's absolute time can be set up. If a line is drawn at right angles to each of the galaxy world lines there is a connection between 'now' in our galaxy and a similar moment in the universal history of every other galaxy. Indeed, a set of such lines (only one is shown in the diagram) defines moments of universal or cosmic time. If this procedure could be agreed upon with people in other galaxies perhaps all clocks could be synchronized to read the same time along each of these lines connecting 'now'. This is very convenient in theory but can only work in practice if the distance of a galaxy and the relative course of its world line are known. With that information, however, that galaxy could be described at a certain epoch in the history of the universe. So the regular expansion of the universe provides a distinct and unique way of referring to definite moments in universal time.

A second feature emerging from Figure 5.4 is that, going back in time, the world lines converge and ultimately meet at one point in space and time. First, however, let me say that it is deceptive to think of space as an empty framework in which events happen. Modern cosmologists define space by the matter in it in the same way that Leibnitz did, and in the early universe, when the galaxies were closer, space was 'smaller'. The universe began at a moment in space/time, but it happened everywhere, not at one place. The whole of space was reduced to a point, there was nothing else. Indeed, in the Big Bang space/time itself was created. By adopting the model of the universe that began with an explosion it is possible to say that the universe has an age. There was a time when it began, and although scientists are a bit uncertain about what happened in the very early stages of the universe because quantum uncertainties come into play as space/time reduces to a very small dimension, the creation itself is a distinct moment from which clocks and calendars can be started. Scientists and cosmologists do not usually draw connections between the Big Bang origin of the universe and the creation in a biblical sense, and indeed may not comment at all on whether the universe was 'created' in the Big Bang. Nevertheless, such a model of the universe corresponds with the Judeo-Christian linear approach to time.

In the first fractions of a second in the history of the universe, quantum effects in theory became dominant and the known laws of physics ceased to work. Big Bang cosmology suggests that space and time became undefined close to time zero. The Big Bang was the creation of space and the creation of time and before that . . . no one can say.

To look out into space is to view only a fraction of the time and distance since the origin of the universe. This is not due to poor instrumentation but because of a feature of the universal expansion itself. The further away a distant galaxy is, the faster its velocity of recession; there comes a point at which a galaxy is so far away that its recessional velocity equals the speed of light. Any galaxies farther away would be invisible to earth and earth to them. The distance at which that boundary or event horizon occurs is the edge of the observable universe. In this sense the observable universe is not altogether unlike a black hole, at least in the sense that light cannot escape out of the event horizon (i.e., galaxies beyond the event horizon cannot be seen). Although it is hard to imagine the universe as a black hole as such, the boundary around any place in it makes that place a special region of space/time; it may be quite difficult to distinguish between the properties of the observable universe and those of a black hole of equivalent mass and volume. Figure 5.5 shows that the light line from our galaxy never reaches the world line of another galaxy once it is sufficiently far away. This diagram also indicates how red-shift occurs as the light from a receding galaxy needs to travel successively farther through expanding space to reach an observer.

If a photon is travelling from A to B through the universe, to that photon its destination is 'known' when it starts out because of its timelessness and its interpenetration through space and time. But during its journey the space expands, and so it too must expand to bridge that journey from A to B. Expansion, in this case, is a longer wavelength. Yet another interpretation of the red-shift, and related closely to this latter one, is that the light must expend some energy to overcome the universal expansion and loss of energy manifests itself directly as lengthening wavelength. So the red-shift is due to universal expansion and is manifest in an altered condition of the light that interconnects with the rest of the universe.

Of course, the history of the universe may not be as we have described it. An alternative theory to the Big Bang describes the counter-balance of universal expansion by continuous creation of matter so that the universe always looks the same. In this 'Steady State' theory new galaxies would form to replace those which had

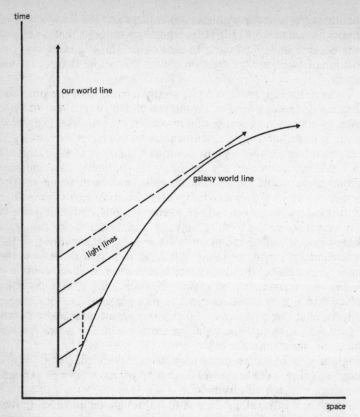

Figure 5.5 *The observable limit to our view of the universe. The light to or from a receding galaxy takes increasingly long to reach us. The increased path from one moment to the next, shown as the thickened line, indicates the redshift. When the galaxy is far enough away our light beam never reaches it and we no longer see it.*

moved beyond our vision. Such a cosmology does not permit an age of the universe, but only of local objects within it. The new matter appearing in the universe is created out of the vacuum fluctuations already discussed and is not a major problem in description. However, in a Steady State universe the concept of distinct epochs becomes rather fuzzy at the edges because sooner or later the apparently similar view of the universe would in fact be made up from galaxies different from those in the past. In this sense there would be not a linear sort of time, but rather an organic cyclic feel to time, with new epochs arising as the region of galaxies on view was replaced by newly created ones. The time scale for such a cyclic renewal would obviously be very long.

The Steady State theory of the universe leads to the prediction that the universe in the future will be just as it is today, expanding uniformly. The Big Bang cosmology leads to different endings. In one alternative, the universe would continue expanding forever. Galaxies receding from each other would become more and more separated until each passed through each other's event horizon. Each would then be effectively isolated from all the other galaxies and would appear to lie in an increasingly cold and empty universe. A form of 'heat death' would occur as the galaxy aged and the stars died. Eventually time would also die because no processes would continue, no light would flow. There would be a dark silence.

However, the universal expansion could slow down, stop and reverse into a universal contraction. The galaxies would move closer to each other, light would abound and the universe would become more concentrated, more dense. Such a scenario would ultimately lead back to the conditions in which the original Big Bang took place, except in reverse. The moment of singularity, when everything was in 'one place', could even lead to a new explosion and another phase in an oscillating or pulsating universe. Time would restart as the next phase of the universal pulsation began.

Of course, in a collapsing universe the density of matter would be constantly increasing as the galaxies all became closer together and, as density increases, time slows down. As the universe approached its ultimate coalescence, time would slow down and cease altogether when the inward collapse was total. This is the inverse of what happened in the Big Bang, when time must also have been running slow. George Gamow, in his description of the Big Bang, talks about the creation of the particles, and atoms and the myriad things that occurred in the first half hour after the Big Bang, and then comments 'nothing much else happened for several million years'. But how long was that first half hour? Is it to be judged by modern clocks, which are running faster than clocks in that very dense stage of universal history? In that case, how is it that so much happened in such a short time interval? If time is marked by change, clocks in the early universe must have been going at a very fast rate, for so much happened so quickly. The solution to this paradox, like solutions to most paradoxes, is to side step the situation to get a fresh angle on it. In this case, remember that space and time are inseparable quantities, connected by the speed of light, and that the scale of the universe in its early stages was quite different from its present scale and should be judged as such. Space and time expand with the universe because they are the universe. The paradox is also one created from the way the situation has been described. If time is 'that quantity measured by a clock' then clocks in the early universe ran slow. If time is defined as change then time in the early universe ran fast as events were changing rapidly. It could be that atomic vibrations occurred more rapidly at the same time that gravitational intensity increased, thereby slowing time, and the two effects then cancelled out. In any case, the limit of understanding is confined by the description used.

All these different cosmologies and the varying consequences of their predictions sound so opposed to each other that it should be easy to be able to distinguish between them. In practice, the different versions all relate to the form of that line connecting the different world lines in the space/time diagram shown in Figure 5.5. The curvature of that line determines the type of universe, either one expanding forever, one expanding uniformly, or one that expands leading to later contraction. Determining through observations the characteristics of that line is a priority in astronomy. But since each galaxy is viewed at a different moment in time, what is seen are different segments of different lines. An alternative method of discovering the nature of the universe depends on finding out its average density, to see whether there is enough material in it to enable it to collapse again; but again, the measurements are not yet sufficiently reliable to reach a specific decision. The trouble is that the universe is so vast and its time scale so long that hardly anything changes during the period over which observations are made. Not only that, but the universe is seen at different stages in its life history. At this point in the history of cosmology, scientists cannot determine the direction of its evolution.

However, universal expansion *does* seem to be related to time's one-way flow. The universe may be considered as a sort of photographic clock: the photos which show galaxies closer together would indicate an earlier time. In later pictures galaxies would be further apart so even if the photos became jumbled up, their correct order could be re-established. The discussion of the gas clock at the beginning of Chapter Four led to the thermodynamic arrow of time and the concept of maximizing entropy in the universe. Likewise, a direction can be given to time due to the expansion of the universe and the recession of the galaxies. Is there a connection between these two strikingly similar cases?

The consequence of the Second Law of Thermodynamics is that all systems tend to thermodynamic equilibrium, that is, to a state of uniform temperature and maximum disorder. If the system considered is the whole universe, then such a state is undergoing what is called the heat-death of the universe. In moving toward such a situation, systems display a directionality in time – the thermodynamic arrow. The heat death, of course, results not only in a uniform temperature, but essentially a cooler temperature. Systems 'run down' in temperature towards the cooler end. Thus, the gas clock's right hand chamber in its initial, empty condition was colder thermodynamically than the filled left chamber. Its cold and empty space acted as a sink for the source of gas molecules in the left chamber to pour into. In the final state, all is homogeneous and uniform in temperature: the drain has been filled and a local heat death has occurred.

In discussing the increase in entropy connected with thermodynamic processes the entropy of the whole universe has been mentioned, because no system can practically be isolated from everything else. This involvement with the whole universe is not just a practical matter, however, because the universe as a whole acts as the sink into which entropy, disorder, flows; into which time flows in one sense. Entropy continues to increase, galaxies continue to recede because the universe is expanding. If that expansion were slowed and put into reverse the thermodynamic arrow would have to reverse as well. The link between the increase in entropy and the expansion of the universe is not causal, but the two arrows of time point in the same direction because of a correspondence between the two phenomena. The universal expansion provides a perfect sink that constantly keeps thermodynamic equilibrium, the heat death, from being reached. Furthermore, in a universe that perpetually expands, space becomes effectively empty when all other galaxies have disappeared over each other's observable horizons. At that stage a different kind of heat death has been reached which imposes constraints on other aspects of time's arrow in relation to different physical events and their interconnection with universal evolution.

Other arrows of time can be linked to the expansion of the universe, and I want to discuss one or two such cases. This book has already pointed to a number of asymmetries observed in physical phenomena. The above two cases are examples, and so is the lack of symmetry between the numbers of particles and anti-particles found in the universe. These asymmetries arise not out of the laws of physics as encapsulated in the standard equations but from other contingencies and boundary conditions. Maxwell's Laws of Electromagnetism are time-symmetrical; time can quite happily be reversed in the equations, which still obtain meaningful answers, except the phenomena predicted are only observed when time runs forward. Consider the case where electromagnetic radiation (light) is generated by oscillating electric charges. Maxwell's equation predicts that a wave of radiation will travel away from the moving source at the speed of light, reaching more distant places at later times and hence is called a retarded wave. If the direction of time is reversed, the equations will predict waves that radiate out from the oscillating charge, reaching more distant places at earlier times. This is, in effect, equivalent to a wave converging in on the source from elsewhere, reaching it at the moment the source oscillates. Such a wave is called an advanced wave, and is pictured in Figure 5.6.

Advanced waves are never observed and indeed their existence would lead to some rather remarkable paradoxes. Consider this example. Imagine that I can generate a signal from which advanced and retarded waves are emitted. The retarded wave will move out through space forward in time. The advanced wave, however, will flow through space backward in time. Now imagine that I have a partner who can receive and send similar signals. He agrees not to signal to me *unless* he receives a signal from me and I agree not to send a signal if I receive one from him at some specified time. When that time arrives I send him a signal. Some time later he picks this up

Figure 5.6   *Advanced and retarded waves from a source, S.*

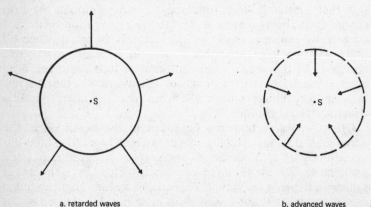

a. retarded waves   b. advanced waves

and replies as agreed. However, the advanced wave part of my signal will have reached him at an earlier time – in fact before I had sent the signal. When he receives that wave he will signal to me, as agreed. The retarded part of his response will reach me at the specified time, when I have agreed not to signal if I receive a wave from him! The rules we have agreed upon have broken down because of the advanced waves moving backwards through time. In such an arrangement an instantaneous response would always be received from an object at any distance away. If it takes a million years for the retarded wave to reach a receiver that receiver will respond, via the advanced wave in minus a million years, that is, instantly. Such a response is called, classically, action at a distance, and was believed in Newton's time to be the way information travelled. However, the speed of light seems to limit transmission times and instantaneous responses are not observed.

Two American physicists, John Wheeler and Richard Feynman, analysed this difficulty with advanced and retarded waves, and the lack of observed symmetry between them. They talked about the response from receivers, which can be at any distance from the source and in any direction, in terms of the response of the rest of the universe (as, they argued, receivers could be found all over the universe). The symmetrical theory predicts, on first inspection at least, an instantaneous response by the universe to a moving charge. Wheeler and Feynman presented a version of the same theory that shows why advanced waves are not observed: the response of the universe, it turned out, cancels them out. It does so in this way. Consider a source surrounded by a sphere, which represents the rest of the universe. The retarded wave setting out from the source reaches a point in the rest of the universe and disturbs that point, which in turn radiates back. So the retarded wave from that point eventually arrives back at the original source in what is regarded as normal communication in the universe.

However, in addition that distant point will radiate backward in time an advanced wave which effectively can be regarded as converging in on that point. At the moment the original source oscillates, generating the signal, the advanced wave from the distant point passes through the source and effectively travels along with the source's outgoing retarded wave. This occurs for all points in the rest of the universe so the retarded signal wave is coincident with the advanced waves converging in on the rest of the universe. In this way, waves seen as travelling out from radiating sources are in practice a combination of the retarded waves from the source plus

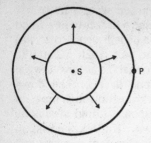

Figure 5.7 *A source radiates a retarded wave, that propagates out towards the rest of the universe. The point P is a place elsewhere in the universe.*

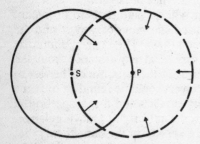

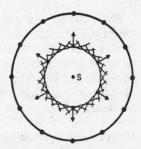

The instant the retarded wave reaches P, it responds by emitting a retarded and an advanced wave. The advanced wave will pass S at the same moment the source wave was emitted.

If all points P in the rest of the universe respond likewise the moment S radiates all the advanced waves cross S and combine with the retarded wave to propagate out from S.

the advanced waves from the rest of the universe (see Figure 5.7). Such a cancelling out occurs only if the rest of the universe is a perfect absorber of radiation. Because it is expanding the universe is a perfect absorber, just as it also acts as an ever increasing thermodynamic sink.

There are now three arrows of time, thermodynamic, cosmological and electromagnetic, and all three are pointing in the same direction and are closely interconnected. Waves travel out from an oscillating electric charge because the universal expansion acts in such a way as to remove advanced waves. Even if advanced waves were permissible entropy would not be increased. Radiation emerging from a moving charge is an irreversible process, more disordered afterwards than before. (The reverse case would be radiation coming in to stop the charge moving.) It is even possible to extend the electromagnetic arrow of time to the atomic or quantum level and say that the reason that electrons in atoms can spontaneously fall to lower energy states with a consequent emission of radiation, but cannot spontaneously jump to higher energy states, is also a consequence of the absorbing quality of the expanding universe.

However, there are problems associated with the ever-expanding universe and its perfect absorption which affect consideration of both advanced/retarded wave asymmetry and the spontaneous emission of radiation from atoms. As has been discussed, a kind of 'heat death' occurs when all the galaxies have receded away beyond their respective event horizons. At this stage there is nothing left in any part of the observable universe to act as a 'response from the rest of the universe', or to act as a perfect absorber of emitted light. At this stage in universal evolution the arrows of time will also cease to operate. Alternatively, as Narlikar and Hoyle have argued, this situation will not arise because it breaks down the idea that physical phenomena are dependent on when they occur. They put forward the argument that such an implication for the future of the universe points in favour of the Steady State cosmology which will always act as a perfect absorber, because new absorbing material is constantly being generated. There has been insufficient response to this proposal from Big Bang cosmologists to enable it to be evaluated critically and the choice between the two outlooks may end up by being subjective.

I would like, now, to mention my own preference between the Big Bang and Steady State theories. It seems to me that one of the difficulties with the Steady State theory is its lack of a beginning, thereby requiring a contemplation of the concept of eternity, in a non-religious context, which does not necessarily feel right. I shall return to this particular theme later in the book and only add here that although eternity as an idea seems to play an important and traditional role in man's consciousness, it does not enter into the Big Bang cosmology at all. Although this cosmology deals with infinities, there seems to me to be a qualitative difference between infinity and eternity, the former of which involves an endless extrapolation from the present while eternity is somehow outside of the present and its continuation. In addition, the organic, cyclic property of the Steady State theory seems to satisfy me emotionally far more than the linearity of the Big Bang cosmology, so I 'irrationally' find it has more appeal. Hoyle and Narlikar's continued support of the theory, and their adaptation of its suppositions to explain newly observed phenomena, I regard with warmth and view the renewal of interest in this and other alternatives to the standard cosmology a healthy sign of the times.

To return to the main theme, there is one last – but important – point about radiation in the universe to consider which involves the symmetry between emission and absorption. Tetrode wrote the following intriguing account of these complementary processes:

The sun would not radiate if it were alone in space and no other bodies could absorb its radiation. If for example I observed through my telescope yesterday evening that star which let us say is 100 light years away, then not only did I know that the light which it allowed to reach my eye was emitted 100 years ago, but also the star or individual atoms of it knew already 100 years ago that I, who then did not even exist, would view it yesterday evening at such and such a time. . . .

Such a story is not altogether surprising, for although a hundred years passed by for the observer, no time at all passed for the photon. Its 'knowledge' of the 'future' is instantaneous. Such an interpenetration of the universe by so basic a phenomenon is one more illustration of the complete interdependence of microcosm and macrocosm, of the atomic and the universal, and the dependence of all events on everything else in the universe. Ernst Mach, the late nineteenth century philosopher and mathematician, related this interconnectedness to the inertia of matter (Mach's Principle).

Mach was intrigued that a free swinging pendulum maintained the direction of its swing in one fixed plane with respect to the distant stars, as the earth rotated beneath it. (This phenomena can be observed at the Science Museum in London, where the swings of a Foucault pendulum mark out the rotation of the earth.) He argued that the pendulum was influenced by the rest of the universe, and his case can be illustrated by a reductive argument. Imagine the earth rotating. Consider then what happens if all the other matter from the universe is removed. When the universe is empty, except for the earth, how can one tell the earth is rotating? What can it rotate with respect to? Unless there is something else in the universe, no motion can exist and hence matter cannot display inertia or resistance to motion. The same argument also applies to radiation. If the sun were the only object in the universe it could not shine as there would be nowhere for the radiation to go.

Rosen has called the idea that all physical properties depend for their existence on everything else in the universe, which implies the interpenetration already discussed, the Extended Mach Principle. In relation to time, this principle would suggest that it is the universe itself that creates time in the same way that it creates space. Space and time do not exist as if they are something absolute (Newtonian) and separate from the rest of the universe, but are produced by it and therefore are relationist (Liebnitzian) and interpenetrative.

This interpenetration of time also connects the universal temporal direction with such 'local' arrows of time as the geological, the biological and the psychological. The continual building of memory constitutes a distinction between past and future which can be seen to be connected with light and space/time and hence to the whole universe and its expansion.

So far, the distinction has not been made between time as it enters the scientist's description of physical reality and other sorts of time experienced. With psychological time, however, it could be that a quite different aspect of time is being discussed from that of the physicists and cosmologists. If time is multifaceted (like light is) then in different disciplines, in different descriptive systems, it may be observed in quite separate, although ultimately united, aspects of temporal manifestation. Beyond the notion that time is connected with change, very little has been explicity said about it. So far, time has been used instead to fit descriptions or explanations of other physical phenomena. I will identify and distinguish some of these descriptions briefly.

The sort of time employed in most physical sciences is called *proper time*. Proper time is time as measured by an ideal clock that is at rest with respect to the observer and is located in a gravity free situation so it indicates some sort of ideal time. The first three minutes of the universe means the first three minutes as indicated by proper time. Those minutes, as indicated by the ideal clock, would be the same sort of minutes as those that pass by as you read this book (which would approximate closely to the minutes indicated by a clock or watch, which is stationary with respect to an observer but is situated in a weak gravitational field).

But if time is considered as change, then the very active early universe changed more rapidly than it does now. Another sort of time, *physical time*, is related to some measure of changing events. Certainly, proper time is also related to changing events but it has been idealized to be the same at all times. Physical time ran a lot faster in the early universe and runs slower now and could be fixed by taking some unit of change as its primary source. Change – and physical time – will cease in the future as the universe reaches its heat death, but proper time will continue forever. Proper time began, according to cosmologists, about fifteen thousand million years ago, but in physical terms the start of time might have been an infinitely long time ago. When did the first change occur?

This problem is not an unfamiliar one. A time system has been divised, which is fine for most purposes, but which has descriptive limitations at the extreme points. The same problem arose in describing light by wavelength or by frequency. The wavelength scale starts at zero and continues to infinity, the frequency scale starts at infinity and continues to zero.

The physicists C.W. Misner and J.W. York have devised rigorous mathematical models of time scales to handle those situations in describing the origin and far future of the universe with a resulting technical time called *York time*. York time starts at time minus infinity at the initial Big Bang and proceeds to time zero when change ceases in the distant future. According to York time the universe is infinitely old but has a finite future whereas in proper time the universe has a finite past but an infinite future. If universal expansion reverses and the universe collapses back to a final cosmic egg, then in York time that will be after an infinite time, although it will be at a finite moment in the future in proper time.

Both of these time systems accurately model physical time in the present and each has its merits at the extreme points in the theoretical history of the universe. Of course, the term *universal time* is based on the uniform expansion of the matter in the universe as deduced from a space/time diagram. In universal time a particular epoch in the history of the universe can be meaningfully discussed, given the assumptions built into such a space/time representation. Universal time is one form of physical time and others are *gravitational time* and *atomic time*. I want to finish up this discussion of time and the universe by examining the relationship between these two manifestations of measured time.

The rate of time keeping with gravitational clocks is not necessarily the same as with atomic clocks. Also, the gravitational constant, G, is the scaling factor in the relationship between the force of gravity, which runs a gravitational clock, and the masses involved. The atomic clock makes use of the electrostatic force which is scaled to the electric charges involved by a property of empty space, called its permittivity, another universal constant. On the scale of a hydrogen atom, the ratio between the electrostatic and gravitational forces is 10^{40}:1, which is an enormous difference, but nevertheless only a difference in scale. The theoretical physicist Paul Dirac, who suggested the idea of the anti-particle, also put forward the idea that this large number had significance in relation to the age of the universe. This is, he suggested, in a slightly different context to this argument, because the ratio between these two fundamental forces is not itself constant in time but changes with universal epoch, and, as the ratio is related to G, the gravitational constant also changes in time.

An alternative approach to Dirac's would be to argue that G is constant but the permittivity of space changed instead, as the universe expanded (remember it is the space itself that is expanding). In this way, the ratio of the electromagnetic to gravitational force could still alter with time so that the 10^{40} large number is that value now. Yet again, both G and the permittivity could both change equally leaving 10^{40} as a fixed value, not changing with time. In this latter case the universe would have only one sort of clock; gravitational or atomic clocks would be related by scale only. However, in either of the former cases gravitational clocks would not in general keep the same time as atomic clocks.

The argument in favour of a changing G, as against the other two possibilities, relates to the notion of the unification of the four forces of nature (adding the strong and weak nuclear forces to gravitational and electromagnetic forces). The weak and electromagnetic forces were unified theoretically by Abdus Salam and Steven Weinberg at the end of the 1960s, but it is the gravitational force that seems most elusive. Theoretical physicists prefer the idea of a single initial interaction which broke into asymmetrical components, the four forces, in the initial stages of the early history of the universe. In that case, the gravitational force would have been much stronger in the past and hence G would have been much larger than its present value.

Indeed, there is some evidence now accumulating that the value of G is changing with time. Records of eclipses of the sun have been used in a retrospective study of the changing relationship between the earth and the moon. The tidal forces of friction between these two objects perpetually wears away some of the rotational energy of the earth, transferring it to the moon, which subsequently moves slightly farther away from us. The eclipse records enable astronomers to work out the rate of this exchange of energy in time. Of course such a process means using the earth/moon system as a clock, and hence the use of gravitational time. In 1978 astronomers in the United States were able to do a comparable study using atomic clocks, finding a significantly different result, which indicated that atomic clocks were running faster than gravitational clocks or that G was decreasing with time, by about the amount predicted by Dirac.

The evidence for this possible variation in G with time is still being tested so the puzzle is not yet resolved. What are the consequences of a changing G for a view of time? It is simply this: if the gravitational constant changes with time then a time interval produced by a gravitational clock will alter with respect to a time interval produced by an atomic clock. If such a situation exists then the age of the universe is not unique but depends on what sort of clock is used. Similarly, its radius will vary depending on what measuring rod is used, which will inevitably involve the use of velocity and hence time and hence gravitational or atomic clocks. A changing 'constant' of gravitation could introduce a new 'relativity' of time: not only would relative motion affect clocks being compared with each other but so too would the epoch in which those clocks existed.

Such a cosmology reduces the status of time, in that time is no longer a single valued or unique parameter in its fabric. But I did start out this work by suggesting time is multifaceted so this description of time need not contradict that principle. Nevertheless I think it does reduce the position of time in its relativism, much as time is reduced for convenience in switching between proper time and York time. So the question might be asked, does cosmology, the discipline that studies the universe as a whole, really deal with the problem of time? A tentative answer would probably be in the negative.

Certainly there are aspects of modern cosmology which explicitly deal with the arrow of time, and this is not surprising as this aspect of time is virtually the only one directly addressed by the physical sciences. Time points, as do other phenomena, toward a view that the universe is interdependent and interpenetrating. This arises largely because of the connecting properties of light, which itself has weaved its way in and out of this analysis. But beyond these two there is little comment on time as such in cosmology.

Time itself has been bypassed by cosmologists in the sense that it has been spatialized. In the spatialization of time the speed of light appears. Time multiplied by c equals space. There is a primacy about light of remarkable character; it seems to contain everything yet seems also to contain nothing. 'In the beginning ... God said

"Let there be light" and there was, and God saw that it was good.'
In the early moments of the Big Bang all was light and after it came
matter. Certainly the problem of time is intricately connected with
light. Measurements of its speed always give the same value, but its
'real' speed is unknowable. Light cannot be observed objectively.
Seeing it is its destruction. It is not by chance that time and light are
bound together, but neither is it certain that a causal connection
exists between them. It cannot be said that time is caused by light or
is the effect of light, but the ideas of causality and chance are bound
up with ideas about time, even with the spatialized time of a space/
time diagram.

If time has any quality or qualities of its own they are masked or
even destroyed by time portrayed as space. The qualities of time
make no sense in the framework of space/time. Time as space is de-
temporalized in a way analogous to the de-humanization that occurs
when human qualities are denied in a quantification of personality.
It is to those qualities of time that I now turn.

6

CAUSALITY AND
CHANCE

The cosmologist Tommy Gold has said about relativity theory that 'all future events to which we may ever have access can be seen, from a suitably moving observer's viewpoint, as simultaneous with the present (or arbitrarily closely so).' He concludes, 'the flow of time is abolished, but the relationship between events is still unchanged.'

Gold is considering the effects of travelling at a speed very close to that of light, thereby connecting the present with the future in an intimate way without losing a principle of causality. Travelling relatively faster or slower does not affect the sequence of events, although it may affect the order in which those events whose occurrence is not determined causally are seen. If a carpenter is hammering in a nail the falling hammer will always be seen first, and then the nail entering into the wood, with the sound of the hammering always following the action. Causal events, by definition, precede effectual events. Nothing in the rather odd world perceptions which emerge from relativity theory contradicts that simple view. (Although the breakdown of 'normal' space/time inside black holes may provide theoretical exceptions to this idea.) Neither is causality threatened in considering the direction of time's arrow in the universe at large. It is only at the fundamental particle level of quantum effects that time reversal and the principle of causality are questioned.

The notion has been put forward that time reversal can occur at the sub-atomic level, or rather in the descriptions of that level of reality, and that the question of whether one thing caused another to occur in that realm is not always meaningful. Nevertheless, the macroscopic outcome of quantum uncertainties is always in the form of time running forward and causality operating. Causally connected events always come in a particular order or sequence, and all observers see them so regardless of their relative positions, motions or ages. By causality I mean, at least, an ordered sequence of events.

A second and fundamental meaning for causality is that one event causes a particular outcome. Nails entering wood do not happen spontaneously but are caused by the hammering. In this way cause and effect are inextricably bound together, so that when a particular effect, such as a rainbow, is noticed, the observer asks himself what caused it. The history of mankind abounds with causal questions. What makes the moon go round the earth? What makes it rain or thunder? What causes flowers to bloom? And the answers to these questions have varied from one culture to another and from one period in history to another. In pre-scientific societies, where man and nature were regarded as both material and symbolic manifestations of a more complex reality, the answers to these causal questions took the form of myths and archetypal stories, with several layers of meaning and significance. In modern, scientific societies the answers are, of course, scientific; testable, predictive answers that lead to practical applications. These answers come in the form of instructions. If you want such an effect, you must provide the following causal circumstances. The form of causal laws is instructional and quantitative rather than qualitative. It is no longer acceptable to think of gods banging the clouds together to cause thunder, but that sort of causal explanation is now replaced by laws of electrostatic discharge. Causality is not just related to the ordered sequence of events that are causally connected, but it lies at the basis of understanding phenomena.

Before turning to the role of chance and of statistical laws of cause, I'd like to consider some problems associated with the 'straight' causal picture developed so far. Consider turning on an electric light. Throwing the switch causes electric current to flow along a circuit, driven (caused) by the voltage difference at either end of the wire, in turn caused by the conversion of fuel into electric potential at the power station. The flowing current causes the lamp filament to heat up by electrical resistance and glow, thereby emitting light. In the atoms of tungsten metal comprising the filament, the electrical and thermal energy causes electrons in the atoms to be excited to higher energy states than before the switch was thrown and in becoming spontaneously de-excited the atoms emit photons of light. So far cause and effect can be traced along the sequence of events quite neatly, but to probe any further involves the quantum world where causality and temporal sequence cease to operate in what are regarded as normal ways. Even in the wire conducting the electricity, cause and effect break down at the subatomic level. It is clear that causal explanation only holds at some sort of macroscopic level and is not necessarily a fundamental property of nature, but rather a useful approximation to what is occurring.

Consider the example of a bus stopping. From the vantage point of a passenger wanting to get off, he was the cause of the bus stopping. He wanted to alight at the cross-roads, and so stood up and rang the bell. Had he not done so the bus would not have stopped. However, from the driver's point of view, this passenger had nothing to do with it; it was he who heard the bell, and decided to act on that signal by following the convention of stopping the bus by applying his foot to the brake pedal. The people waiting at the bus stop think that it was they who caused the bus to stop because they signalled to it to do so, although they would concede it was the driver who reacted to their signal. It may turn out that this bus stop is one at which all buses must stop, so neither the passenger's signal nor that of the people waiting at the stop had any say in it.

The cause of the effect in this example is relative and multifaceted. In principle all the causes mentioned, and others too, combined to provide an overall cause resulting in the bus stopping, but having moved from a purely 'physical' example to an example that includes human action, there is an additional complexity in trying to determine a cause or causes. Even in this case, an attempt to push the physical braking part of the picture down to a subatomic level with its attendant statistical, apparently acausal connections would be superfluous.

Another example of finding a meaning for causality could be by asking the question of how two cars crashed into each other. The first solution to the problem seems obvious. The driver of the red car drove his vehicle out of a side turning without due care and attention and hit the blue car that was passing by. That caused the accident. But did it really? The driver of the blue car could have passed that point seconds earlier or later and not been hit. If he had not overtaken the slow lorry half a mile back he would not have been hit. If he hadn't forgotten his briefcase when he went to the garage and had had to return to the kitchen for it he would not have been there at the 'right' moment; and similarly for the driver who 'caused' the accident. A whole string of actions that delayed or speeded up both drivers' journeys acted by chance to cause them to converge on one place at one time. The lack of care and attention was caused perhaps over uncertainity about the time of a meeting, because the telephone line had crackled when instructions were being communicated. Perhaps the lack of care and attention was aggravated by a fly that had by chance flown into the car and was distracting the driver. Again whole chains of events each acting as a partial cause farther back along myriad converging paths of cause and effect led to the accident. It is a case of one damn thing after another, like time itself!

In such an analysis, it is very hard to pin down a cause as the chain of linked events will lead eventually back toward a 'prime cause' in an infinite regression. Perhaps the point to notice in such an exposition is that the links in the chain of reasoning are of dual nature. Some links are causal, strictly, while others are only causal by chance. For example, the car moved forward because the driver put his foot down on the accelerator, which caused petrol to be ignited in the cylinders etc, and the indecision or lack of care was caused by a fly entering the vehicle by chance. It could be argued that there are two quite different ways of describing the causal chain, rather than two types of link in the chain. One chain could be described purely quantitatively while the other is qualitative, but I think it is more sensible to see the two types of description as interacting as suggested above. In this case all the chance encounters, delays and so forth 'averaged out' in such a way that the blue and red cars arrived at the same place at the same time. The overall effect of chance events, the random nature of some of the causes, was to provide an accident, which can be regarded as the macroscopic outcome of microscopic incidents; or the improbable outcome of a series of possible events.

My final example relating causality more firmly to chance and probability comes from Warren Weaver's book *Lady Luck*.

The circumstances which result in a dog's biting a person seriously enough so that the matter gets reported to the health authorities would seem to be complex and unpredictable indeed. In New York City, in the year 1955, there were, on average, 75.3 reports per day to the Department of Health of bitings of people. In 1956 the corresponding figure was 73.6, in 1957 it was 73.5. In 1958 and 1959 the figures were 74.5 and 72.6.

In this example the effects are caused by dogs biting people, because they were hungry, aggressive, protective, mad, and so forth. Each event was independent of the others and it can be presumed that no dog knew how many of its fellow kind had bitten people on any one day. Here random events show an overall pattern which allows statistical and probabilistic causes to events to be postulated. On average, seventy-three to seventy-five dogs were likely to bite people on any one day in New York over a five year period. The limitation of such statistical information, however, could be shown in the situation where a Department of Health official, having just heard of the seventy-sixth case of a dog bite on that day, was cornered on going home by a mad Alsatian in a closed alleyway. He would know the improbability of his being bitten but would that knowledge make him ignore the seriousness of his predicament? Statistics unfortunately do not help in the individual case.

These examples cover a number of ways of thinking about causality and the nature of causal laws, and indicate a number of problems associated with the idea of causality. On the one hand, cause and effect go together well in the macroscopic world and provide a useful means of describing the origin of effects and sets of instructions for producing effects; but at the microscopic level causality breaks down due to quantum uncertainties. Then, on the other hand, where causes are part of chains of events the element of chance enters the picture and alters the nature of causality in a different way, although the handling of both chance and of quantum effects involves the use of statistics. In fact, when it comes to describing events, causal laws governing the occurrence of the events cannot be separated from statistical laws applying to the same phenomena. Causality and chance are opposite sides of the same coin. They are both characteristics of nature and are both related to time in their fundamental meanings.

By chance I mean the unpredictable, as opposed to the predictability of causality. The unpredictable can also be described by simple laws, the laws of chance. The laws of chance and the mathematical understanding of the ideas concerned with probabilities were devised originally in the field of gaming, but it soon became apparent that not just the unpredictable behaviour of rolling dice or spinning roulette wheels are subject to laws of probability, but so too could any set of unpredictable events, such as dogs biting people. Whether such an extension of the statistical laws is totally valid will be discussed in due course, but from the dog biting example it is clear that similar unpredictable events have a gross, statistical pattern to them. So too does the distribution of characteristics in a normal population; for example, the height of adult people in a population will cluster around the average height with fewer and fewer exceptions at the extremes of the height distribution. Such a characteristic of a population is based on the averaging process inherent in handling large numbers of individual cases, and the heights of a small group of people may not fit into such a smooth distribution. As the population increases, providing it is not being selected with any bias, the actual height distribution will approximate closer and closer to the ideal curve.

The laws of chance and probability, expressed by statistics, relate, then, to the average properties of a group, whether the group be a series of throws of a pair of dice or of daily rainfall or of the ages at death of racing car drivers. Such laws say what is likely to occur on average or how rare a particular event is. For example, in a series of runs of ten tosses of a coin, ten heads in a row will only occur once every 1000 runs. On most days about seventy-four dogs will bite people in New York, although rarely it may be as few as fifty or as many as a hundred. Statistical laws indicate the odds to be expected in future events, based both on past experience and theoretical expectation, and point out the rarity of an individual event after it has occurred. If the statistical law is describing a characteristic of a population then the real population can be compared to that ideal to see how typical it is.

Statistical descriptions began to be used by physical scientists in the eighteenth and early nineteenth centuries to describe the gross properties of matter, and are seen, for example, in the gas laws, where the pressure and temperature of gases of various volumes are described. In these cases statistics are really being applied in a significant way because the assumption (required for statistical meaning) that the individual events being treated are all identical or identically equivalent is taken to be true (one gas molecule is just like any other). Also, in any real sample of gas there are countless numbers of particles, maybe 10^{20} or more, so the average properties of the particles will dominate totally in the sample as a whole. Ideally, one assumes an infinite population but 10^{20} is so close to infinity that it makes no difference. The success of the gas laws and other macroscopic descriptions in physical science illustrate the success of the statistical method, but it does so because the population is large and its members essentially identical.

Statistical laws, then, describe the behaviour of the unpredictable, and do so in a quantitative fashion that has its own measure of exactness. The laws of chance complement the laws of causality and both sorts of law are found in science. The quantum world is unpredictable, so that the laws of quantum theory are statistical. The motion of an electron, for example, is described only in terms of the probability of its location. When particles are described as waves of probability, they are reduced to mathematical abstractions. Quantum reality, by the nature of its description, contains inherent uncertainty and unpredictability and so displays acausal properties, which are expressed mathematically by statistical methods. Quantum theory lies at the root of modern physical science, so science must of necessity combine causal and acausal modes of operation in its descriptive system.

Causal laws always lead to prediction, by stating what effect is produced by a given set of causes, but statistical laws are non-predictive at the individual level although they contain a predictive element about the average properties of a population. The unpredictable, the acausal, compensates for when causality cannot hold. The complementary character of these two sides to phenomena provides the counter-argument to determinism. Laplace suggested that given the positions and speeds of all particles he could predict the future completely, at least in principle. Because macroscopic and sub-microscopic areas of nature are both unpredictable in detail future courses of events cannot be determined and neither can the present conditions.

One last comment about statistics and chance is that although statistics may reveal overall or average properties of a population, looked at in a particular way, no inference about a common cause, or any cause for that matter, can be drawn from the statistical significance. The fact that seventy-four dogs bite people in New York on average each day tells us nothing qualitative about the incidents. Indeed, the statistics may lead one to infer a supposed connection linking the incidents, where none exists. The events are single, individual and truly unpredictable. In this sense statistics associated with smoking and death from cancer of the lung do not establish a causal connection, although it is quite clear that smoking leads to a high possibility of dying prematurely from cancer or other smoking-related disease. It is a known and quantifiable risk, just as New Yorkers expose themselves to a one in 400 chance of being bitten by a dog in any one year. The smokers who live well into old age are compensated for, in the averaging process expressed in the statistics, by those who die very young. With statistics, remember, the individual case is not dealt with, it always remains unpredictable.

The primacy of order is maintained even when chance is considered. At least at the macroscopic level chance can be the unpredictable cause of an event; there is a characteristic ordering of events, and it is this order which remains an inviolable principle in relativity theory. When time stands still, however, like inside a black hole, then causality also ceases. No time means that before and after cannot be defined, so causality ceases to apply. If in travelling round inside a rotating black hole I meet an image of myself from the last time around how do I know it was the last time and not the next time? Worse still, how do I know which one is really me?

When time is reversed causality also suffers. If a tachyon could interact with matter then it would do so before it had even begun its journey. Anti-particles are effects followed by causes, in one sense of the description, as can be seen in some of the examples in Chapter Four. To return to the behaviour of electrons in 'causing' light to be emitted from a lamp, once the vacuum level is reached what is going on becomes clouded with uncertainty as causality becomes confused with a form of chance – the myriad and unpredictable ways particles can manifest themselves. In turning to smaller and smaller time intervals, a level is reached where causality is no longer distinguishable in nature. This temporal view of the quantum world leads to the insight that causality is itself a temporal phenomenon which only appears at longer time intervals. Causality is a principle that nature displays at macroscopic levels of time.

There is another way to look at this finding. Consider the problem of action at a distance, discussed in the last chapter. Action at a distance has always worried scientists because of the lack of a mechanism by which it can operate. For example, how could the force of gravity work between the sun and the earth without some means of transmission? Field theory offers one solution, whereby the properties of space are altered by the presence of the bodies concerned, such that objects placed in the field respond as if a force was operating on them. Einstein's gravitational field theory describes space as being distorted or curved by the presence of massive bodies, thereby explaining the forces observed. Furthermore, changes in the body giving rise to the field would alter the field, with the alteration 'radiating' out across the field at the speed of light. In reality, no action at a distance is involved. But the example of advanced and retarded electromagnetic waves showed that the effect of the advanced waves (travelling back in time) is elminated by the response of the rest of the universe in such a way that their effect was instantaneous but concealed. In other words, action at a distance operates if advanced as well as retarded waves are permitted. If electromagnetic waves propagate in this way, gravity should also, in which case Mach's Principle would also be a case of action at a distance. The problem of action at a distance is that it eliminates time, and hence causality, and that, rather than its lack of suitable mechanism, is, I suspect, the origin of its unacceptability to modern science.

Action at a distance is instantaneous and therefore cause and effect happen at the same time, which is a meaningless statement in relation to an operational definition of time, as simultaneity is not verifiable for events that do not occur at the same place. However, action at a distance is simultaneous by definition, as advanced and retarded waves show. This effect was one part of the evidence for the idea of the interpenetrative nature of the universe, and this sort of interpenetration only works when events can occur simultaneously, with instant response, without time and without what is normally called causality.

So, in addition to the dependence of causality on the large scale manifestation of time, there is also a lack of causality in the timeless concept of action at a distance; just as causality does not have meaning to a photon of light. Depending on how physical reality is described, events are either ordered in a causal sequence, unfolding in time, or occur simultaneously in a timeless and causeless world. (I use the word 'causeless' here only in the sense of absence of temporal sequence.)

It appears at this point that causality as a fundamental principle is being squeezed out, that ultimately everything can be reduced to chance, to unpredictability, and hence to purely statistical treatment. This is certainly the direction in which modern physics is moving. Quantum theory, which is *the* fundamental theory in physics and the basis of all current reductionist views of science, allows statements to be made such as that physical properties at the sub-atomic level are completely lawless and are never capable of being related by any kinds of causal laws. Such an attitude toward the basis of physical reality goes hand in hand with the wider view that the universe is utterly meaningless, governed only by chance and necessity.

David Bohm, in his book *Wholeness and the Implicate Order*, questions this random view by showing how hidden causes, not allowed by orthodox theoreticians, can be reconciled with the fundamental quantum theory. His exposition demonstrates the plausibility that quantum effects are manifestations of a deeper, implicate order in nature, not due to chance; that what seems random in one reference frame is not disarrayed at that more fundamental level. Things merely *seem* random and causeless. It is as if I drew my name on the page edges of a closed book; on each individual page there would appear a random mark or two of ink with no apparent cause or pattern. It would only be by closing the book and seeing the ink marks from another perspective that the cause would reveal itself.

A physical example of a random, lawless and causeless phenomenon is that of radioactive decay, when an unstable atomic nucleus gains stability by ejecting an electron or an alpha particle. Radioactive decay is cited as the prime example of a purely and intrinsically random process because the time at which any individual atomic nucleus will decay is completely unpredictable. But radioactive decay, though a chance event, can also be described by the laws of probability. Indeed, because atomic nucleii are so similar and because there are so many of them in any realistic sample of, say, uranium, the statistical description of radioactive decay is very precise. What can be predicted is the time it takes for half the given sample to decay, the well known half life, which is an average and well demonstrated property of radioactivity. At the level of the individual nucleus, however, there is no way of predicting whether it will decay now, in a second, in a minute, a year or even for millions of years. Not only is there no means of predicting it, but it is held as a part of scientific belief that it is incapable of ever being predictable. There are no hidden causes.

In practice, radioactive decay is placed in a special category, quite apart from other unpredictable events; it is described as causeless, absolutely. Furthermore, the random decay is uninfluenced by any physical or chemical processes occurring to the sample. Scientists can heat up, cool down, and bathe in acid radioactive material but its decay remains as random as ever and the half-life is unchanged. In contrast, the occurrence of dog bites in New York are unpredictable but nevertheless have causes, and are subject to environmental effects (*viz* the pattern is different in Los Angeles). The non-predictableness of radioactive decay is truly acausal, although in general non-predictability does not necessarily imply causality. At least the random decay is *said* to be truly acausal, but hidden causes can be postulated quite legitimately and have been, in fact. If hidden causes were discovered to underlie radioactive decay then the random nature of the decay would be viewed as previous ignorance. Similarly, if environmental factors were shown to affect the decay rate so too would the principle of true randomness be violated.

Professor Horace Dudley has suggested that the cause that triggers the effect of radioactive decay is the interaction between particles in the nucleus with those massless, chargeless and elusive sub-atomic particles, the neutrinos. Neutrinos flood the universe with their presence. Millions of them pass through every object each second, and their lack of mass and charge makes them invisible even to solid rock. No mechanism is known whereby neutrinos might trigger radioactive decay but it is at least an hypothesis which would knock the complete arbitrariness of radioactive decay off its special pedestal. Of course, the neutrinos which bathe the universe are randomly scattered throughout space (not unlike gas molecules in a room) so the suggested interaction would still be unpredictable, just as it is for dogs biting people in New York – but it would not be causeless.

This sort of suggestion has, however, been received with hostility by the scientific community at large. I myself stand, with Professors Bohm and Dudley, on the minority side of the fence in that I share with them the view that there can be hidden causes and that the apparently inherent randomness of radioactive decay, for example, may be a measure of ignorance.*

Randomness itself can be viewed as a relative and not an absolute phenomenon. Like quantum theory, randomness can be shown to contain pattern or meaning from another view point, from a deeper order implied by the perceived manifestation. The possibility of fundamental and hidden causes allows some underlying meaning in an apparently causeless universe.

Everything that occurs can be regarded as an effect and therefore as having a cause, in the sense of 'what made it happen'. Saying that something happened 'by chance' is only to say that its cause is unknown. The event may not be predictable but something must have made it happen, and causality in this sense must surely be a fundamental property of nature, even in the cases where cause and effect are simultaneous and even at the sub-atomic level.

So if causality can be regarded as that which made it happen, what is meant by acausality? Consider those dog bite statistics again. Each individual case obviously had its own cause, its own history, but the statistic actually refers to the average property of the larger population and nothing causes seventy-four dogs each day to bite people. The statistics refer to an acausal property of a collection of events, not to the individual events themselves, and no causal agency is operating to bring about this average. There are no causal

* Dudley's suggestion that the neutrinos form a kind of ether which might trigger radioactive decay questioned the orthodoxy of the time in the 1960s. In the present climate of opinion his suggestion has stimulated useful and rewarding research into the neutrino. His ideas, like some of those of Hoyle, might not be right but they have activated research in fruitful directions.

connections between the individual incidents. I said above that unpredictable events are not necessarily acausal in the way that predictable events are causally connected (hence their predictability). The unpredictable dog biting incidents are causal in that something made the dogs bite people, but the collection of incidents altogether is acausal. The incidents correspond to an overall pattern, which appears well ordered, although acausal.

Such correspondences are quite common in everyday experience, wherein one set of events either displays what appears to be a significant pattern or else matches with another set of events in such a way that a meaningful correlation is noted. An example of the latter type of correspondence is the often noted one of the height of the hemline of ladies skirts with the state of the stock market. The apparent correlation does not imply a causal connection.

Another type of correspondence occurs between a symbol and the reality for which it stands. For example, the symbolic representations of hills, rivers and roads on a map correlates to the actual geographical feature in the landscape; but the connection between the two is acausal. A building does not cause a model of itself to be made, or vice versa, but there will be a correspondence between the two objects, even though the correspondence can never be exact. This sort of acausal connection is certainly not non-predictable or due to chance, but demonstrates meaning and purpose. Chance and necessity do not govern this sort of acausality, and yet it lies at the base of much experience.

One sort of correspondence already encountered in a different guise is that between an actual event or series of events and the expectation of such events occurring. That is, there is a correlation between real events and theoretical probabilities, and understanding and acceptance of much present-day knowledge is based on the faith placed on such statistical correspondences. When a torrential downpour in mid-July is described as being very rare, the actual rainfall is being compared with the rainfall statistics for that time of year. If my friend buys four lottery tickets and wins a prize on each one of them then I can calculate the odds of her fortunate windfall; but I do so by comparing the actual event with its chance likelihood. The idea of chance likelihood comes from probability theory, which itself makes use of the concept of randomness.

Consider 10,000 lottery tickets printed, with 1000 winning numbers distributed at random. These tickets are well shuffled and distributed to different vendors, so that no one can tell which individual ticket is a winner. The chance of any one ticket being a winner is one in ten. This argument is based entirely on the idea that randomly shuffled items are acausally related and that the chances are evenly distributed among the items. Normally, one would expect to have to buy ten tickets to obtain one prize; and to get four prizes out of four is remarkable, although the odds are one in 10,000 – so it may well happen once in a batch of 10,000 tickets. It is nevertheless a rare and lucky occurrence. If my friend goes and buys another four tickets and wins on all of them I will start to wonder if she is more than lucky, since the odds for that are one hundred million to one.

But perhaps real events are not random in the purely theoretical sense? There are two ways to approach this question, either through considering the appropriateness of using the concept of randomness in a given situation – that is, are statistics always the right approach? Secondly, what is meant by randomness itself?

The idea of randomness developed alongside the idea of probability and the laws of chance which presume that no causal agency is meant to be playing a part in the outcome. One result is as likely as any other, providing the mechanism concerned is fair. Randomness is a perfectly reasonable idea and one associated with our modern western culture. Indeed, the attachment of randomness to scientific culture is why it seems reasonable, because culturally reason stands very highly among human faculties. However, all things being equal, there seems to be no conceivable reason why a coin should fall either heads or tails. Each toss of the coin is considered an essentially identical operation, each roll of the dice is like any other, every spin of the roulette wheel is just the same as any other spin. If this assumption is true, then, the chance of any one result appearing is a non-predictable and hence random event. Random often means inability to predict, but the laws of chance compensate for this form of ignorance.

Can mathematical techniques be related to events which are not strictly identical and have unrelated causes? Do the statistics about dog bites in New York really have any significance? Just because events display average behaviour does not necessarily signify anything other than the fact that if an average of some data is taken then that data displays an average. Reducing complex events to statistical numbers in fact detracts from the significance of the incidents and only displays rather limited if not useless information.

When statistics and probability theory are used to predict expectations of chance events the idea of randomness also enters, and this concept itself is a theoretical notion. Much of the following analysis has been influenced by George Spenser-Brown's incomparable little book *Probability and Inference in Scientific Investigation*. His is the only complete discussion I know in which the limitations of the expectations of chance are examined in such depth. Random events are those which display no discernable pattern, are truly unpredictable, and appear to be without purpose or cause. Statistical laws, rather than causal laws, determine their outcome (although 'leaving things to chance' sounds as if chance is a manifestation of a hidden cause or causes!).

The idea of randomness should also be applied to a *series* of events as well as to individual events that are unpredictable. True randomness applies to a series that displays no discernable pattern in addition to being a series made up from unpredictable items. As it is to the series as a whole that randomness is applied, randomness itself then is a concept applied to something already produced. In other words, a single event can be random if its occurrence cannot be predicted in advance, but a random series always involves comparison of the series produced with a model of randomness in hindsight. A random series should show no discernable pattern, and if one is perceived then the random nature of the series is denied. However, the inability to discern a pattern is no guarantee of true randomness, but only a limitation of the ability to see a pattern. For example, a pattern in a string of random digits may not be discernable in the first several thousand checked, but may be clear on a larger scale; after ten thousand digits the sequence may repeat itself, thereby becoming perfectly predictable. Checking a sequence for randomness is very difficult.

One problem with random events is the question whether the sample looked at is the whole population or merely a sample. Imagine programming a computer to produce a sequence of random digits and stopping it after one hundred digits have been issued. One would expect to find roughly ten ones, ten twos, etc. In practice, there would be variations; there may be only eight sixes but twelve nines, for example. However if there were fifty ones it would be most alarming and would lead one to the conclusion that the computer was not producing true random numbers. (In fact, any sequence of random numbers produced by a computation is only a pseudo-random sequence. It may show no discernable pattern but it is produced by a formula, and hence is predictable in practice, even if it otherwise matches very closely to an ideal random number generator.) However, if the computer produces a string of one million random numbers, out of which one subgroup of a hundred digits produces an excess of ones, it would not be out of the ordinary. Indeed there would be a reasonable probability of such an occurrence happening, and if the sequence produced ten million digits such an excess of ones could be expected at least once in any group of one hundred digits. The initial alarm was due to an unlikely event occurring in the first subgroup which had been mistaken for the whole population.

In principle, the whole population may be an infinite string of random digits, in which case anything that was produced would be possible, so even if one thousand ones appeared in succession, it would not be too surprising. Randomness is very much in the eye of the beholder, as what may pass for randomness in one context may be quite useless in another. A series of ones and noughts may appear quite random for use as a sequence against which to compare the tossing of a coin, heads equals one, tails nought, but it also might be the binary code version of a well known song and therefore perfectly predictable and full of pattern to someone familiar with binary notation.

Is it possible to have a perfect random number generator? Many attempts have been made to produce such a generator or its equivalent by publishing tables of pseudo-random numbers. The producers of such tables hope they have spotted any possible patterns and corrected for any initial discrepancies, but they can never be entirely sure. The thing that is regarded as truly random, and which can therefore be employed as a random number generator, is radioactive decay. By taking a mass of radioactive substance and counting the emitted decay particles with a suitable detector, such as a Geiger-Muller tube, the irregular spacing, or time intervals, between emissions can act as a randomizer. Since a sequence of events can be random only by inspection, not by prediction, can radio-activity be validly used as a model for randomness? The individual events of a decaying nucleus are unpredictable in time and the substance as a whole represents a series of such events, many of which series have been inspected, so it appears that it could be a good model of randomness. But if a causal mechanism is ever found, the series would become intrinsically predictable and hence pseudo-random. The purposelessness of random behaviour would have been destroyed, and only its lack of regular pattern preserved.

There is an even stronger reason for doubting the true randomness of radioactive decay and that comes from an experiment, performed by Walter Levy and Eve Andre at the Institute for Parapsychology in the U.S. in 1970. In this experiment the effect of the relative lengths of light to darkness on the behaviour of young chicks was studied. An automatic light switch was programmed to be on and off for twelve hours a day in total but the length of each period of light or darkness was governed by a radioactive decay controlled randomizer. The apparatus was checked carefully and confirmed to operate correctly, the light being switched on and off at random times but with equal amounts of lightness and dark over twenty-four hours. The chicks were placed in the experimental set up and it was found that the periods of light always exceed the total length of darkness in any day. The chicks preferred the light to be on, it made them warmer and more agreeable generally, and their wish somehow or other affected the randomizer, that is, the rate of radioactive decay. This finding was confirmed in a second experiment Levy conducted using unborn chicks in fertilized eggs, which excercised the same 'control' over their environment, although sterile eggs produced no such deviations from the expected arrangement.*

This experiment casts great doubt over the true randomness of radioactive decay quite apart from any other questions it raises. But before jumping to any definite conclusions about mind over matter, the expectation of the true randomness of the experimental set up should not be forgotten. It could be that the effect observed in the above case is an example of the unusual happening when a theoretically random result was expected. It often appears that initial results of experiments do not confirm prior expectations because reality does not operate according to theoretical calculations. In addition, there is the problem of the random series.

Any series may appear random, until someone discovers the pattern, or conversely, a series may not appear random because the expectations of what a random series should be like has not actually been produced. Such a sequence may then be tampered with until it looks more random, but is it? If the series has been 'massaged' to appear random then how can it really be? Purpose has been applied to make it appear 'without purpose'. The real problem of randomness is the inability to truly test for the quality sought. As Spenser-Brown has said '. . . the absence of one pattern (in a random series) logically demands the presence of another. It is a mathematical contradiction to say that a series has no pattern.' But if a preconceived concept of what randomness should be is continually fed back into a randomizer then the series produced will not be truly random but will only model an idea of what such a series should be.

* I have been unable to discover the reproducability of this experiment or whether it has ever been replicated.

Tests of randomness only test theoretical expectatons and not the real series at all.

Spenser-Brown suggested that 'chance expectations' were not realistic because of the failure to randomize in practice; and that the problems already discussed would lead one to expect non-standard results in an experiment where, instead of comparing guesses with a random series, one random series was compared with another. Mr Arthur Oram performed 'An Experiment with Random Numbers' that appeared to confirm Spenser-Brown's view, but later he discovered that his own organization of the material gave rise to the high deviation from expectation and that, if he treated the same data in different ways, then different outcomes and less significant ones resulted.

To overcome the ambiguity of Oram's experiment Robert Harvie performed a similar experiment using computer generated series of pseudo-random numbers. Harvie's experiment demonstrated that there were consistently too few correct 'guesses' when one random series was matched against another. Indeed in around fifty thousand attempts the number of correct guesses fell short of the expected result by so much that it was a statistically less probable occurrence than one chance in over a thousand.

To my knowledge the experiment has not been repeated since 1973 and so I investigated this property of random expectation myself. Having access to several computers ranging from desk top minis to a large main frame machine I generated several million random numbers and made extensive tests for their obvious randomicity and for the success rate of one series 'guessing' another. In the first kind of experiment I compared pairs of pseudo-random digits generated successively by a computer. The results are summarized in Table 6.1a. It is clear immediately that the guesses were not as successful as chance expectation so that after 175,000 guesses on three different computers, deviation from expectation was nearly 500. The probability for such a discrepancy is 0.000043 or one in 23000; high odds indeed.

In some ways the scheme was not unlike having two packs of cards and turning up cards simultaneously from each pack in a game of 'snap'. In case there were obvious hidden patterns in the sequence being generated I then 'shuffled' in between each comparison in this way. Two strings of eleven random digits were generated, and the last digit in each string examined thus:

Guess	Target
53742099671	37286985046

Table 6.1a *Results of comparing pairs of pseudo-random digits (0 to 9) and recording the number of 'hits', i.e. when both numbers are the same. MCE stands for mean chance expectation, deviation is the deviation from that expectation and CR is the critical ratio (deviation divided by the standard deviation) telling how probable that deviation is. A CR of 1.65 corresponds to a probability of 0.05 or one chance in 20.*

No. of comparisons	MCE	Result	Deviation	CR
5000	500	480	−20	0.94
5000	500	476	−24	1.13
5000	500	458	−42	1.98
5000	500	486	−14	0.66
5000	500	488	−12	0.57
5000	500	488	−12	0.57
5000	500	488	−12	0.57
5000	500	501	+ 1	0.05
5000	500	485	−15	0.71
5000	500	481	−19	0.90
5000	500	478	−22	1.03
5000	500	510	+10	0.47
5000	500	467	−33	1.56
5000	500	471	−29	1.37
5000	500	483	−17	0.80
25000	2500	2292	−208	4.38
10000	1000	979	−21	0.07
25000	2500	2611	+111	2.34
25000	2500	2506	+ 6	0.13
25000	2500	2358	−142	2.99

total number of comparisons = 175000
number of hits = 17007
MCE = 17500
deviation = −493
CR = 3.93
probability = 0.000043 or 1 in 23000

For the guess the last digit is 1, so the first digit of the string is selected as the guess; it is the number five. Similarly, for the target number the sixth digit is nine, and so, in this attempt the guess is wrong. This randomizing the random sequence should get over any obvious patterns produced by the algorithm generating the pseudo-

Table 6.1b *Results of comparing pairs of randomized pseudo-random numbers in terms of the number of 'hits' found.*

No. of comparisons	MCE	Result	Deviation	CR
5000	500	470	−30	1.41
5000	500	456	−44	2.07
5000	500	465	−35	1.65
5000	500	477	−23	1.08
5000	500	473	−27	1.27
5000	500	440	−60	2.82
5000	500	438	−62	2.92
5000	500	465	−35	1.65
5000	500	457	−43	2.03
5000	500	472	−28	1.32
5000	500	548	+48	2.26
5000	500	542	+42	1.98
5000	500	532	+32	1.51
5000	500	504	+ 4	0.19
5000	500	485	−15	0.17
10000	1000	1068	+68	2.27

total number of comparisons = 85000
number of hits = 8292
MCE = 8500
deviations = −208
CR = 2.38
probability = 0.0087 or 1 in 115

Table 6.2 *Summary of author's experiments combined with those of Harvie.*

	No. of comparisons	Result	Deviation	CR
Experiment 1	175000	17007	−493	3.93
Experiment 2	85000	8292	−208	2.38
	260000	25299	−701	4.58
Harvie's Ex. 1	24800	2364	−116	2.46
Ex. 2	24800	2385	− 95	2.01
	49600	4749	−211	3.16

random digits. The results of these experiments are also given in Table 6.1b. Here a total of 85,000 comparisons were made (which means generating nearly two million random numbers) and again too few guesses were successful, giving a deviation from the chance expectation of one in 115. This result is not so dramatic but still compares with the findings of Robert Harvie. Table 6.2 shows the combined results of my experiments with Harvie's. Clearly, Spenser-Brown's argument is borne out.

What do these results demonstrate? One thing they show is that pseudo-random number generators do not produce truly random series. But how can one tell what is random in practice? Results like these led Spenser-Brown to conclude that 'they [the results] comprise, in fact, the most prominent empirical reason for beginning to doubt the universal applicability of classical frequency probability.' That does not mean, of course, that statistical findings should not be trusted, but it does mean that extra careful handling of results is necessary when a comparison with chance expectation is made. Like time, randomness is something believed to be understood until it is examined and really tested.

Another thing demonstrated is the relativity of randomness; I quote Spenser-Brown: 'the concept of randomness bears meaning only in relation to the observer; if two observers habitually look for different kinds of pattern they are bound to disagree upon the series which they call random.' Randomness is in the eye of the beholder, it is not found truly in nature, which implies not only that nature is patterned, but also that the idea of randomness is really an expression of ignorance. If randomness is disposed of, what, then, can be made of chance?

As David Bohm has said, chance is necessary in nature to complement causality. And yet there is reason to doubt the primacy of causality. Causality and chance are only ways of describing what is believed known and what remains unknown. Laws of causality and laws of chance can be constructed but they are not fundamental, they are still only descriptions. Perhaps more significant are the acausal correspondences which occur time and time again as this investigation of time continues, for this diversion into causality and chance is here to provide a foundation from which to view other aspects of time.

The relativity of randomness, the view of chance as a way of handling ignorance, leads to a final question. Is nothing due purely to chance? If by chance is meant the inability to predict, that is, ignorance, then the answer could be ambiguous because no one knows. But it can be surmised that if even such a random thing as radioactive decay is not really random, and if it is agreed that everything must have a cause, in the sense of something that made it happen, then the answer must be resoundingly no, nothing does happen by chance; to everything there is a purpose. The universe is not random, events are not governed by chance and necessity, but to all things there are meanings and prime causes. Those meanings and causes are only named chance. It is not chance, then, that the speed of light has the value it has. Time itself does not appear by chance; like all manifestations in the universe it has meaning and purpose, which is why time has such a richness and quality that chance can never have. Some of that richness can be seen in a wide variety of human temporal experience. The coming together of time and randomness and chance is most clearly seen in the experience of coincidence.

7

COINCIDENCE AND
MEANING

If two events happen at the same time and place the occurrence is described as simultaneous, although constrictions must be placed on the description of events for simultaneity to be maintained. But when two events occur at the same time and place and the two events are identical or closely similar and are not causally connected then their simultaneity is called a coincidence.

If I am talking to a friend and he asks me whatever happened to a mutual friend, Bob, and I tell him I haven't heard from Bob for over five years, then it would strike me as a coincidence if the telephone rang at that moment with Bob phoning up 'out of the blue'. Such an occurrence would be a coincidence because its probability was sufficiently small to make it distinctly notable. It would have barely been noticed if my friend had enquired about my wife, who was out shopping, had she at that moment turned up at the door. It can be seen at once that coincidence has something to do with probability and chance.

However, unlikely events occur all the time. I may be having a talk with a friend, whom I rarely see and who surprisingly mentions Bob whom I did not know that he knew, when the phone rings with a call from someone else who contacts me with far less frequency than does Bob. The improbability of the simultaneous events may be more extreme than in the case where Bob was the caller, but it is not recognized as a coincidence, only as an unlikely set of more or less simultaneous events. Coincidence involves simultaneous events of a closely similar nature or events that have a relationship additional to and separate from their individual causes and their simultaneity. In other words, a coincidence can be recognized because of the parallel meaning of the events. If, when I am out with a friend, we call in at a house belonging to an acquaintance of his but who is a stranger to me, and the phone rings with someone asking for me, not knowing I am there but having dialled a wrong number, the event would be regarded as an amazing coincidence.

But if the call is just a wrong number the event has no meaning for me and, although it may be just as improbable, I would not regard it as a coincidence. Coincidence, then, seems to involve a coming together in time of unlikely events which are connected by a separate but significant meaning to the person experiencing it. A coincidence is merely an improbable simultaneity unless it is experienced by someone to whom it is significant, not just because of his presence but because he has brought to the occurrence a special meaning. In this sense a coincidence can be thought of as being a triple simultaneity requiring the presence of one event, the person experiencing the event and a significant meaning connecting the first two. Of course there may be more than one event or there may be a sequence of events that go to form the coincidence, but it is essential to have an experiencer and a connecting meaningfulness.

The experience of coincidence and the attachment of meaning to the events that makes them form a coincidence are properties of consciousness and of being human. It must also be recognized that coincidences must always remain anecdotal in that they are both individual and personal. A coincidence happens to someone. It is a one-off event and only significant to the person to whom it occurs, although it may well appear meaningful to others on hearing an account of it, but that is because they share with the person to whom the coincidence occurred an account of the meaning or significance of the otherwise neutral events. Because coincidences are anecdotal, personal and individual, they are not open to scientific examination or statistical analysis. It is almost impossible to think of quantifying coincidences even in the way New York dog biting can be quantified, for although the biting incidents are not causally connected they have enough in common with each other to withstand a statistical comparison. Coincidences are seldom comparable, are all quite distinct and therefore must remain anecdotal, outside scientific evaluation.

Coincidences occur in many different ways. The sort involving telephone calls or visits I have already used as examples. The astronomer and philosopher Camille Flammarion collected many anecdotes of this nature, with perhaps the story of Monsieur de Fortgibu as the most endearing.

A certain M. Deschamps, when a boy in Orleans, was once given a piece of plum pudding by a M. de Fortgibu. Ten years later he discovered another plum pudding in a Paris restaurant, and asked if he could have a piece. It turned out, however, that the plum pudding was already ordered – by M. de Fortgibu. Many years afterwards M. Deschamps was invited to partake of a plum pudding as a special rarity. While he was eating it he remarked that the only thing lacking was M. de Fortgibu. At that moment the door opened and an old, old man in the last stages of disorientation walked in: M. de Fortgibu, who had got hold of the wrong address and burst in on the party by mistake.

This multiple coincidence gains its weight from the repetition of simultaneous events. The second coincidence of pudding and M. de Fortgibu is itself not so remarkable, especially as the man seems very partial to the pudding. The final episode, however, is especially dramatic as Fortgibu only arrives at the event, to make it a coincidence, by accident (dare I say by chance?).

Authors and researchers often expeience coincidence in the form of references or key information falling into their hands at the right moment, sometimes quite literally. Colin Wilson has described how in trying to remember from where a particular quotation came was amazed when a book fell from the shelf and lay open on the floor at the passage he was seeking. Arthur Koestler quotes a case reported by Dame Rebecca West, who was searching in Chatham House for a record of an incident in the Nuremburg trials:

I looked up the trials in the library and was horrified to find they are published in a form almost useless to the researcher. They are abstracts, and are catalogued under arbitrary headings. After hours of search I went along the line of shelves to an assistant librarian and said: 'I can't find it, there's no clue, it may be in any of these volumes.' (There are shelves of them.) I put my hand on one volume and took it out and carelessly looked at it, and it was not only the right volume, but I had opened it at the right page.

An incident that occurred to me involved a double search. As an astronomer with an interest in astrology I was looking for a fellow astronomer with the same unlikely interest. One afternoon while pursuing astronomical research I needed to look up an obscure Czechoslovakian reference, in a journal I had never even heard of. I went to the library and pulled open the appropriate catalogue card index drawer and the cards flopped over of their own accord revealing the index card for the journal I was seeking. The journal itself was on a high shelf, but conveniently the ladder was in exactly the right spot for me to climb up to the book. As I climbed I couldn't help but notice a new astronomy book peering out between the rungs of the ladder and which I hadn't previously noticed. I removed it from the shelf, still perched on the steps, and flipped it open. There on the page was the horoscope for the author of the book. He too had more than a passing interest in astrology.

Quite trivial things can attract coincidences. A friend and his family were driving along a main road when one of the children said: 'I wonder what happened to our old car?', referring to a vehicle that they had sold over two years previously. They recalled its registration number and then, looking out of the window, saw it coming along the road toward them.

There is a well-known phenomenon, common to everyone, whereby when a certain detail or specific is brought into consciousness, then awareness of other cases of the same sort is heightened. For example, on holiday I hired a car of a make I would not normally notice with any regard on passing in the street. But on return from holiday I became very much aware of those cars down every street. However, this increased awareness is not the same as coincidence. Talking about a specific car and then seeing it is, provided the talking was genuinely before the seeing and the likelihood of seeing it remote. And yet the phenomenon of increased awareness is often brought out as a counter-argument to refute the notion of real coincidence.

It is frequently said by critics who do not want to accept coincidences as real phenomena that the so-called coincidence is only noticed because of the increased awareness that one event has set up for the subsequent event of the coincidence. Furthermore, it is argued that all the occasions in which a coincidence did not occur must be taken into account in realizing that its happening was one of those seemingly rare events that are bound to happen, just by the law of averages. Both parts to this argument really miss the point as well as denying what is nearly always an experience of some moment.

I have already described a coincidence as the simultaneous (or nearly so) occurrence of two or more causally unconnected events that are significantly linked by the person to whom the coincidence occurs as being meaningful. The events come together in time, share a moment of time which is imbued with a particular quality that has meaning to the receiver of the experience who is, in a way, caught up in that moment of time. If two events come together in time then the argument about increased awareness is misplaced. If I am talking about Bob when he telephones me, I do not notice the coincidence because I was especially aware of Bob, due to my talking about him. The coincidence was noticed because it was he who then telephoned. The 'increased awareness' was the coincidence. Certainly it was not increased awareness of M. de Fortgibu that made the old man blunder into the pudding party at the opportune moment. The events coincided in time and place and were meaningfully connected by the relater of the incident.

There are certain forms of coincidence that may be excusably explained away by increased awareness which will be discussed shortly, but the sort of coincidences I am describing here are not amenable to such a method of discrediting. The exception could be the sighting of the car, in that it could be argued that the car had been seen by the family on other occasions but not noticed because they were talking about something else. This argument can be dealt with in two ways. Firstly, it must be stated that the point of this coincidence is that the coinciding events were the discussion of the car and its appearance. Any hypothetical occasions when the car might have been spotted are irrelevant to this case, partly because they may never have happened, which will never be known, and partly because this particular case alone was the coincidence. To discount an experience on the basis of hypothetical 'other experiences' is ridiculous.

The second response to the 'other occasions' argument involves the wider issue of the probability of coincidences and the argument that a coincidence ceases to be 'special' when the times coincidences did not happen are taken into account. There are both philosophical and pragmatic problems with this argument. The pragmatic difficulty is quite simply about the evaluation of the probability for the coincidence and the probability for the 'not-coincidence' cases. If the car spotting example is taken, an evaluation must be made of how often the family drive in the car, how many cars are passed per unit of time; the total number of cars on the road must be worked out, the frequency with which the family drive along the stretch of road where the coincidence took place, likewise the frequency with which the other car passes along that road and so on. Then, how often the old car is mentioned must be evaluated and probably several other factors as well. This sort of information is just not available. To do the job properly it should also be calculated how often similar coincidences almost happened. For example, was there an occasion when the old car was mentioned and it was just round the corner out of sight? This sort of information could never practically be obtained, at least certainly not in retrospect. So a true probability for the coincidence cannot actually be determined. Even worse, what if this was the first time since selling the old car that it has been mentioned at all? The event is then unique and cannot be quantified sensibly.

This last point leads to the philosophical problem with the probability argument. Any coincidence must be a unique event, in a set or class of one item only. It is so essentially by definition. How can any coincidence then be quantified in a probability estimate? Indeed, the question can be asked how, statistically, the probability of any unique event occurring can be evaluated. What, for example, is the likelihood that just as I write this sentence a particular car should draw up in the street outside, the wind flow in a particular direction, and a particular sparrow fly past the window? The combination of all these events occurring at one and the same time and place is quite improbable and yet they did so. There is nothing remarkable about it. But if an attempt was made to realistically evaluate the probability or chance that they would happen as they did, a figure would be obtained of astronomical unlikelihood. There may be cars drawing up in the street ouside every five minutes or so, but for that particular car to arrive at that particular minute (or second . . .?) might be incalculable. How often has that car parked in this street, at this spot in this street, at this time? How precisely is the wind direction to be quantified, to the nearest degree? How frequently do I sit here writing? How often do I write this particular sentence? The improbability could be so large for such an ordinary 'coincidence' that no one would believe it, and yet it happened. Furthermore many, many other unconnected events took place at the same time and place, hence making the betting odds even more ridiculous.

In dealing with unique events in hindsight there is no sensible or meaningful way of estimating their likelihood. Statistics are inappropriate. Coincidences are unique events and cannot be reduced to a probability. At best it can be said that if one notable coincidence is experienced every five years, and that an average coincidence takes five minutes and an average lifetime is seventy years, then there is a one in 525,600 chance of a coincidence occurring in any five minute interval of someone's life. Such a figure is, of course, meaningless.

Probability theory and statistics grew out of games and gambling and then expanded to encompass any phenomena that were either random or showed repetitious patterns. Natural laws are statistical laws because types of events repeat themselves. But just as there is causality so too is there chance or acausality. Just as there is the patterned and ordered so too is there the irregular, disordered and unique. The existence of scientific law implies the existence of things outside those laws. Events can be described either as part of a patterned course of existence or as unique and individual. These two ways of approaching reality reflect two ways in which science can operate and I want to digress briefly to examine them.

The first of these is what I shall call *descriptive science*, following the analysis of Schumacher. Descriptive science is concerned with the search for the whole truth about a thing of interest. There is never a limit to an examination of a subject and nothing found is ever superfluous, because everything discovered, if the examination is honest and truthful, will be a part of the whole truth. In descriptive science the question always being asked is: 'What do I actually encounter?' Such a question is all-embracing and leads to an attempt of careful description. There is nothing unsophisticated or simplistic about this approach to science. Indeed its difficulty and enormous demands, coupled with its limitless scope, make it a fearsome path to set off along, and it is a path that few even attempt. Descriptive science is not just the beginning stages of fact gathering, as for example, in botany or astronomy, as a preparation for the more 'mature' sciences of molecular biology and astrophysics. It is a whole approach of the scientist to the phenomena within and around him.

The second way of thinking about science is what I shall call *instructional science*. In this form the scientist is seeking the answer to the question: 'How can I explain this phenomenon?' In seeking an explanation the whole truth of the matter is not necessarily required, because an explanation may be valid even if it is only partially true. Of course the explanations for a set of observations will be constantly refined, overhauled and even completely replaced as the demands for the explanation to encompass more effects increase. The way in which modern science* explains things is through the mathematical modelling of the phenomenon under investigation. Such models come predominantly in the form of differential equations which act as sets of instructions and hence this mode of science is called instructional. The equations instruct what the effect will be for a given set of causes, initial conditions and so on. Their validity can be tested by following the instructions to see if they work. If they do, or do so nearly enough, then it is said that the instructions explain the thing being modelled.

Instructional science is a very powerful technique for abstracting certain principles from the physical world and putting them to use under different and controlled circumstances. For example, the essentials of flight can be extracted from a study of the wings of birds

* By modern science I mean the science that developed in northern and western Europe from the seventeenth century onwards.

which, when combined with other studies that provide suitable power sources, can enable aeroplanes to be made. However, in every application of this technique the whole truth is not sought. Because the quantitative aspects of phenomena are usually the keys to writing instructions the qualitative aspects are neglected. It matters not at all that the soaring of a seagull is graceful or that the flight of a crow appears cumbersome. What can be extracted from both is a 'principle' of aerodynamics, which can then be used. Descriptive science does not necessarily lead to explanation, and it does not necessarily lead to usage. It was really when the first modern scientists, and their patrons, realized how useful explanatory science was that it predominated.

Instructional science proceeds by finding general laws, such as Newton's Laws of Motion, which provide idealized instructions for producing effects. As such instructions are necessarily causally based so the acausal side of nature is left out of an instructional explanation of phenomena. Time is also squeezed out of the world picture provided by instructional science because, although time enters as a static quantity in some of the equations, its essence is removed and so it is never described or explained. Time is like the acausal and is more often found to be essentially qualitative not quantitative. Time, like other non-causally connected aspects of the world, is not subject to generalized instruction or to statistical description. So it is not found explicitly in instructional science and a descriptive approach is the only valid avenue open to the scientist who wants to examine time.

In returning to coincidence it is plain that instructional science will be found wanting, as such a science cannot describe a single event in terms of general principles. It does not make sense to have a 'Law of Seeing Your Old Car Just When You Happen to be Talking About It'. An instruction cannot be written for a coincidence and to that extent coincidence cannot be explained.

This argument can be pushed even further and the world can be described principally in terms of its unique events, and all events *are* unique. They happen at one moment in space and time, such that an instruction cannot be written for those events' existence. An instruction can be written for a property of events or for a type of event but not for such and such an event to happen at this time and place, where other events are also coincidentally occurring. If the world is described in detail in terms of all its countless simultaneous events, all its 'natural coincidences', then the world cannot be explained.

The causal connections in the world can be explained in terms of natural or scientific laws but the acausal connections cannot be explained nor even made sense of in a statistical or scientific way. It is this fact that makes the study of coincidence so difficult and is why so many people like to discount coincidences as just due to lucky chance. Instead, a framework must be found wherein coincidences, as temporal experiences, lead to an understanding of time and events more clearly, wherein can be found a meaningful explanation that will not be 'scientific'.

So the response to the argument that coincidences are the events that are noticed against a background of many non-coincidences and therefore just due to chance, is to point out that in dealing with the simultaneity of acausally connected events normal probability calculus is inappropriate. Furthermore using that approach leads to the conclusion that all events are improbable which is clearly not the case, for 'events' are occurring perpetually all round everyone! Besides, there is also the question of chance itself. Chance always implies that the unlikely is being compared against a model of supposedly random events. Such a supposition has already been demonstrated to be dangerous and there may not even be any such real thing as randomness, just as infinity is a purely abstract concept. So to dismiss coincidence as merely due to chance, or as 'one of those things' is to ignore the real problems of randomness, to appeal inappropriately to probability theory, and to discount the idea of a purposeful universe.

If coincidence could be considered 'due to chance' in the sense discussed in Chapter Six then the phenomenon should be considered as purposeful. This idea of purpose is enhanced by the same probability arguments that were employed to dismiss the critics who would like to discredit coincidences. If, in describing the world as a series of unique events that are frequently acausally connected, in that they happen to occur at the same place and at the same time, it should be realized that there is a high improbability to their simultaneity, then it might well be argued that so many improbable events actually happening could only occur for a purpose. As the world consists of myriad such 'natural coincidences' at all times and in all places then it might be concluded that the world is only sustained by a grand purpose. Such a discussion then touches on those theological arguments about the proofs for the existence of God. I do not intend to embark on theological discussion here but will remark that this line of reasoning seems to confirm the ancient notion that there is indeed 'a time to every purpose under the heaven'.

The unfamiliarity of discussing nature as acausal, improbable and full of purposeful coincidence lies in the dominance of the notion of causality that is inherent in instructional science. It is just as valid, however, to describe all the unconnected events of the world as it is to describe those events that are causally connected. Such a description is just as 'true'. Its only drawback (or maybe its advantage) lies in the fact that it is far less useful. Descriptive science can combine the acausal with quality or meaning, but the looseness of such a description makes it appear useless. The inability of instructional science to handle coincidences, combined with their acausal uselessness, is perhaps why the notion of coincidences is anathema to so many scientists. A coincidence is a reminder of the forgotten acausal side of things. That, in one sense only, is its purpose. As there are at least as many unconnected or acausal events as there are causal ones, a coincidence can be seen as pointing to this missing half of reality. If a scientist believes that his work is to explain all of reality (and reference to reality can be confined only to physical reality without any loss of meaning) then a coincidence is not a very encouraging reminder. He can dismiss it quite easily as being anecdotal, etc, his mind can be set at ease and the coincidence has, to all extents and purposes, been explained away. Or has it?

I have used the term 'natural coincidence' for events brought together acausally at one time and place which are both improbable and possibly purposeful, but the coincidences I started out by describing are distinct from these 'natural coincidences' in that they have an overt and significant meaning to the individual experiencing the events. The personal perception of a coincidence is an experience which is always notable, usually curious and frequently numinous. One obvious purpose to such a phenomenon is to draw attention to the acausal. The linear, causal view of time is so familiar that, if for nothing else, a coincidence will jolt someone out of that complacency into recognizing the multifaceted, enigmatic and complex nature of time's reality. Coincidence as a temporal phenomenon is a wonderful example of time's duality. If the concept of the dual nature of light as both wave and particle is acceptable, then coincidences should help in the acceptance of the no more strange idea that time displays the duality of the connected, linear, causal side of its nature and its acausal, unconnected aspect. The experience of time through coincidence points to a much more complex, much more bewildering and awesome aspect of nature that was overlooked in the more familiar descriptions of the apparently explicable and seemingly controllable world given by instructional science.

An ideal example of time's duality is encapsulated in the anecdote about a stopped watch given by J.W. Dunne in his book *An Experiment with Time*. Dunne was staying at an hotel and one night he dreamed that he was having an argument with a waiter about the time. He was claiming it was 4:30 in the afternoon but the waiter maintained it was 4:30 in the night. In the dream Dunne realized that he must have been arguing because his watch must have stopped. He took it from his waistcoat pocket and it had indeed stopped – with the hands indicating half past four.

At that point he woke up and the dream had such an effect on him that he lit a match to see if his watch really had stopped. He wrote: 'To my surprise it (the watch) was not, as it usually is, by my bedside. I got out of bed, hunted round, and found it lying on the chest of drawers. Sure enough it had stopped, and the hands stood at half-past four.' It is obvious what had happened. He had unconsciously noticed that the watch had stopped when he went to bed and this fact just emerged during his dreaming. Silly incidentals from one's daily life frequently do that, don't they? Dunne wound up the watch and went back to sleep. In the morning he rose, dressed and picked up his watch. Going downstairs he made for the nearest clock in order to set his watch to the right time. 'To my amazement I found that the hands had lost only some two or three minutes – about the amount of time which had elapsed between my waking from the dream and rewinding the watch!'

The twist in the story is remarkable. Had the watch actually stopped at 4:30 the previous afternoon then he would have noticed it much sooner and put it right. Indeed he claims that it must have been working when he went to bed. That means that it must have stopped at the time he had the dream. Maybe the sudden silence, the lack of that background ticking, caused him to dream of a stopped watch. Perhaps that is the explanation. If so, then how did he know, in the dream, that it was 4:30? Dunne suggests it was a form of clairvoyance, but it seems to me to be more of a remarkable coincidence. The story has several layers to it, both in its telling and in trying to unravel its possible explanation. The duality of time's linear flow, the passing of clock time, is dramatically juxtaposed with the coming together in time of dream, watch and significance in an acausal correspondence. There is no easy explanation for this incident but in its description something is indicated about the quality of time.

The causal imperative is a strong one and somehow always seems to creep back into explanations of coincidence. Often several 'other causes' may be invoked to explain coincidences, such as extra-sensory perception, telepathy and so forth. It can be immediately seen, however, that causes such as these are possibly as far-fetched as acausality itself. Even if such effects occur, which many scientists strongly doubt, the mechanism by which they work is unknown. It can be argued that the conversation about the old car was prompted by an extrasensory perception of its close proximity. Its arrival then ceases to be a coincidence but a natural extension of the hidden perception. My discussion of Bob just as he telephoned me 'out of the blue' could be explained as a case of telepathy. Even the flopping over of the index cards in the library catalogue or Colin Wilson's falling book could be accounted for as cases of psychokinesis! Rather more absurd would be to describe M. de Fortgibu's dramatic entrance at the pudding party as a case of telepathy or ESP, but even in that case the coincidence would still not be explained away. The repeated coincidence of pudding and Fortgibu coming together three times under different and improbable circumstances seems pushing ESP rather too far. It seems so unnecessary to invoke other causes to explain coincidences when they can just be accepted as strange concatenations of simultaneous events or a form of temporal experience. J.W. Dunne falls into this trap with the marvellous experiences and temporal experiments described in his book. Instead of listening to what the experiences are telling him, he tries to squeeze a theory out of them, and, of course, his theory of serial time falls into problems of infinite regression and is far less interesting than the anecdotes he is trying to account for.

I am not, at this point, discounting the wide and varied realm of phenomena called the paranormal, which includes telepathy, ESP, precognition and many other effects. Indeed, in Chapter Nine this whole region of experience will be sifted to see what aspects of the paranormal are concerned with time, and what may be learned about time from those experiences. I am discounting ESP and other such causal explanations in relation to coincidence because it seems to me that the point of experiences such as coincidence can be so easily missed in turning to some device which explains them away. Instead I am asking that the phenomenon is faced and challenged. For by facing it its validity is acknowledged and the lesson it has to teach can be learned. I have argued the case for a purposive physical world, and if ways can be found to let nature speak in order to reveal some of its purpose, far more will be learned about it, however strange and mysterious it is, than will be the case if it is viewed through limited but neat mental and theoretical constructs.

Two scientists who have confronted the problem of coincidences in this century and acknowledged their acausal nature were the Austrian biologist Paul Kammerer and the psychologist Carl Gustav Jung. Both men produced theories about coincidence and both propounded the view that coincidences were phenomena which obey an acausal principle rather than causal laws.

Kammerer's theory was published in 1919 under the title of 'Das Gesetz der Serie', 'The Law of Seriality', and Jung's thinking on the subject appears in his essay 'Synchronicity – An Acausal Principle', which was published in 1952, alongside an essay by the physicist Wolfgang Pauli who worked with Jung on the problems of acausality. Arthur Koestler must be praised for bringing Kammerer's life and work to general notice, especially as the study on seriality has never been translated into English. Indeed, in *The Roots of Coincidence* Koestler examines both these acausal theories in the light of modern physics and research into ESP. This is not the place for a critique of Koestler's book, which, while taking a different path to my own, does provide a useful and informative examination of coincidence. What I want to do here is examine both the theories of seriality and of synchronicity in the context of this discussion of time.

Kammerer's Law of Seriality, which has nothing whatever to do with Dunne's Theory of Serial Time, is based on the observation of series of coincidences. Kammerer noted, as have many before and since, that events often come in sudden bursts, a series of coincidences, a run of luck. Kammerer's seriality refers to cases such as the same number coming up in several different settings, on a bus ticket, a theatre ticket, a cloakroom ticket and a telephone number for example, or the mention of the same place by several different people during the course of one day. Such a series may be found in almost any set of otherwise apparently random sequences of events and can be treated itself as one form of coincidence and hence can be placed into an acausal perspective. The following example was given to me by a friend.

The house which I am buying and which is being built at the moment has until now been known as Plot 8, Courtlands, The Phelps, and the builder was certain that this would be the final address. That is not to be so, as the Council have decided that the numbering of the new houses should continue as The Phelps and my house will be number 44.

It just so happens that my parents live in a new house, built by the same builder, which is also number 44 in its road.

About a year ago my brother decided to buy himself a small flat in London and it was number 23 in the block. (My present address is number 23.) But negotiations fell through at the last moment; but he heard that another flat in the same block might also be for sale. This turned out to be the case and he has now purchased it. It is number 44!

Such a set of obviously unconnected coincidences is typical of the effect that Kammerer observed and described in his own definition of seriality as the '... lawful recurrence of the same or similar things and events ...' which '... are not connected by the same active cause.'

There are two difficulties with this concept of seriality. The first refers back to the earlier discussion about increased awareness for noticing coincidences. A series of unconnected events that appear like a coincidence may always be subject to the possibly valid criticism that the noticeability of the series, the reason why it stands out as somehow special, is just because it does stand out against an otherwise apparently random background. That is, the second member of the series is noticed because it arouses an increased awareness of the memory of the first item in the series. Number seventeen on a bus ticket stands out when the date is also the seventeenth, hence one may be subconsciously looking out for other seventeens, or at least be more aware of that number whenever else it turns up. This argument is very difficult to counter without making extensive observations and measurements of the kinds of series under consideration. Kammerer apparently did just that and half his book is concerned with the cataloging and classification of such runs of coincidences. However, analysis of the results of such careful observations does not reveal the acausal principle which Kammerer regarded as universally operating, but the foundation of the second difficulty that faces his whole theory.

It has been earlier noted that a random series is purely a theoretical construct. Found within a supposedly random series are clusterings and groupings that, although surprising, arise from the very nature of what randomness must be like. If, in a random series of noughts and ones, we notice twenty noughts in succession we understand not that the series is systematic but that such a clustering is quite normal, of the nature of 'randomness'. What Kammerer is noting is surely just such a deviation from what is expected to happen in a random series. Events of a similar nature, such as noticing numbers on tickets, etc, that occur during a normal daily routine stand out because those daily events are regarded as being randomly shuffled. It is not surprising, then, when a series of like events is sometimes found in this random chain. Both these arguments are quite valid criticisms of serial coincidences, but neither argument explains a case like that of M. de Fortgibu, or of the house numbers. The trouble with Kammerer's work lies, I think, in his mistaken juxtaposition of what he recognized in the world as its acausal component and the rather trivial series of events that cluster in pseudo-meaningful ways.

This criticism is like that presented about coincidences in general, namely, that one does not notice all the events that do not stand out. In this case the argument may often be valid in that what is noticed seems like improbable clusterings because the incorrect assumption is made that such clusterings should not occur in a random sequence. But the argument is limited and reduces to a matter, perhaps, of mere classification. If a series is nothing more than a local clustering effect, as most of Kammerer's cases seem to have been, then they should not really be regarded as coincidences at all, especially if the definition that a coincidence is a meaningful coming together in time is maintained. This definition should not exclude series of events that do form a coincidence. It may well be that a clustering type of series is also imbued with a significant meaning which elevates it from being just a curious series to a true coincidence. It is not just odd, but strangely meaningful, that on the day of Prince Charles' and Lady Diana's royal wedding (a day, incidentally, with strong collective feelings throughout British society) that a quarter of the races were won by horses with names such as Tender King, Favoured Lady and Wedded Bliss. Of the 200 or so horses that ran that day eleven had 'royal' or other suitable names and of those six won or came second out of seventeen races. The betting odds on these horses combined to give a 54,000 to one chance against such an occurrence and not one was a favourite! This is a typical clustering and one that is not particularly unusual, and yet it stands out because of its additional meaning and connections with the Royal occasion and the emotions attendant.

It is not surprising that Kammerer used the word 'lawful' in his definition of seriality. It looks as if there could well be some 'acausal law' that operates to strike up these moments of coincidence. However, the problem of an acausal law is very difficult to countenance in a description of nature when that description is essentially made up exclusively of causal relationships. How could an acausal law work? What mechanism is used? These sorts of questions come to mind automatically and are, of course, causal questions and therefore inappropriate to the very notion of an acausal world. Kammerer's definition also uses the phrase 'not connected by the same active cause' which modifies and protects his controversial position on acausality. By 'active cause' he presumably means that a recognizable causal agent is not at work in creating the series, but the word 'active' allows causality, surely, to creep back in by some unspecified 'passive' mode. His phrasing does allow some ambiguity and suggests his confidence in an acausal principle was not without its uncertainties.

Why Kammerer perhaps missed giving a more satisfactory account was by neglecting both the psychological factor which seems to be so closely attached to coincidences and the element of time and its qualitative significance. The presence of meaningfulness in events coming together in time seems to be the crucial test between what can be regarded as a coincidence or a simultaneity. Jung's principle of synchronicity combines a psychological factor and time's quality with external coincidental events.

Jung, like Kammerer, was especially interested in coincidences in his life and those of his patients, but his principle of synchronicity concerns not all coincidences but a special sub-set within the whole gamut of such experiences. I have used the term coincidence to mean those happenings which both come together in time and contain a significant meaning. Jung's synchronicity refers only to those coincidences where a strong psychological state becomes mirrored in an external event: 'Synchronicity therefore means the simultaneous occurrence of a certain psychic state with one or more external events which appear as meaningful parallels to the momentary subjective state.'

The case of M. de Fortgibu and plum pudding, which Jung himself quotes in a footnote, does not qualify as synchronistic, and neither do many examples other people have quoted as synchronicities, which are of course just coincidences. The kind of thing Jung had in mind is best described by the key example he gave himself.

A young woman I was treating had, at a critical moment, a dream in which she was given a golden scarab. While she was telling me this dream I sat with my back to the closed window. Suddenly I heard a noise behind me, like a gentle tapping. I turned round and saw a flying insect knocking against the window pane from outside. I opened the window and caught the creature in the air as it flew in. It was the nearest analogy to a golden scarab that one finds in our latitudes, a scarabaeid beetle, the common rose-chafter (Cetonia aurata), which contrary to its usual habits had evidently felt an urge to get into a dark room at this particular moment. I must admit that nothing like it ever happened to me before or since, and that the dream of the patient has remained unique in my experience.

The beetle, like Fortgibu, arrived at the right moment in time. But where the old man's entrance coincided with another external event, although a significant and meaningful one, the scarab's appearance coincided with a strong psychic state. It is this special internal-external mirroring that Jung describes as synchronicity. However, having coined a term for a special and an especially strong form of coincidence, Jung then elevates the term to 'designate a hypothetical factor equal in rank to causality as a principle of explanation'. Certainly an acausal principle is just as valid a descriptive system as is the more familiar causal principle. Unfortunately Jung has excluded all other aspects of acausality from his coined term and thereby made 'synchronicity' a difficult principle to grasp. It seems to me that synchronicity is not a principle of explanation but a term to encompass a certain range of temporal experiences that can be more generally bracketed under the term coincidence. Indeed Jung himself keeps referring to synchronicity as *meaningful* coincidence, although his definition is much more constrictive than that. As Koestler has also pointed out, the Principle of Synchronicity is as difficult as Kammerer's Law of Seriality. The notion of a law or principle invariably involves some sort of mechanism and hence becomes a pseudo-causal description. Both Jung and Kammerer recognized an acausal side to nature and its powerful connections with meaning and significance but neither could disentangle a 'causal' mentality from the acausal things they were trying to explain. Explanations themselves are causally oriented (instructional science) and Jung and Kammerer would have possibly achieved a more satisfactory stage in their exploration had they been content to describe and not try to explain.

Having said that, synchronicities do occur and are powerful experiences. A remarkable series of synchronicities form the framework of Laurens van der Post's book *A Mantis Carol*. At the start van der Post is working on his book *The Heart of the Hunter*, which concerns the author's exploration of the Kalahari desert and his encounters with the Bushmen of that region. While writing about the instinct that he says had prevented him from ever finding coincidences accidental, Post received a letter from New York sent by one Martha Jaeger. The surname means hunter and Martha was the mother of sons condemned to darkness, metaphorically the mother of the Bushmen. Martha Jaeger was enquiring about the praying mantis, the insect god of the Bushmen, that had been appearing in her dreams. The coming together in time of hunter/ Jaeger, book and Bushmen, together with the theme and significance of mantis initiated a series of synchronicities, of meaningful coincidences, that linked mantis, Bushmen and van der Post in a journey to the United States and an encounter with a deceased Kalahari Bushman in New York. It is a beautiful story which illustrates the difference between coincidence and synchronicity, even though this is a difference largely of degree.

There will always be some confusion over the use of the term synchronicity, probably because a lack of clarity in Jung's own mind about the phenomenon. Certainly, though, a synchronistic experience, combining the powerful inner subjective state with a parallel material event with a closely similar meaning at a moment in time, is a distinctly different form of coincidence from the examples quoted earlier in this chapter, where the main components of the coincidence are the coming together of external events that have a meaning to the observer, but not arising, at least overtly, from his or her subjective state. If coincidences can be regarded as acting as signposts to acausal aspects of reality, and of directing people's attention to the complexities of time, then synchronicities can also be regarded as bringing to notice the importance of a psychic state. Synchronicities may even point out to us that we are psychic creatures and not just material, biochemical, causally comprehended beings. If nature so contrives itself to produce such remarkable parallels to human internal states then such coincidences must surely be saying something about the human condition and its relation to the acausal, which is so often regarded as mysterious.

A third form of coincidence happens when an external event strikes a correspondence with a particular psychic or psychological state, a sort of synchronicity in reverse. External events, such as the sudden shattering of a piece of china, quite unexpectedly, or the sight of some unusual and yet strangely meaningful object, can instil an abrupt awareness of a frequently inexplicable and usually hidden significance. An example that comes to my mind was the sighting by myself and a small group of friends of a pink flamingo standing in a flooded field in rural Sussex. The event was certainly unexpected and a flamingo is hardly the sort of bird one expects to see in the English countryside but the incident held some extra component of oddity about it that made it significant to us all.

This sort of coincidence is a well recognized phenomenon and, although not commonly discussed today, goes by the name, omen. By considering omens as a special form of coincidence, synchronicities can also be put into a clearer perspective. Omens are signs that operate at a psychological or psychic level in much the same way as do synchronicities, except that the external manifestation triggers the coincidence. Whereas with synchronicities the coming together of psychic state and external manifestation has a significance and meaning that indicates something special, so too do omens have a message and a significance.

The reading of omens, the interpretation of signs, is the art of divination about which I shall have more to say in Chapter Nine. Let me point out here that divination is a process deeply embedded in time, as it contains not just an element of reading the portents for here and now but also for the past and the future. Of course, omens seen as a particular form of coincidence are themselves temporal phenomena because they involve the coming together in time of external event and the inner, subjective state aroused by the omen. Furthermore, the external event itself represents something of the quality of time at that moment. The connecting thread for omens, synchronicities and for all coincidences is their placing in time and the fact that all such events are significant because of their inherent meaning.

Interestingly enough Jung himself presents as his second example of a synchronicity an anecdote actually concerning an omen. In this case a woman told him that at the deaths of both her grandmother and her mother, birds had gathered outside the window of the room in which the deaths occurred. The woman's husband was sent by Jung for a medical check-up as he showed some symptoms of what might have been heart disease. The man's doctor, however, could find nothing wrong with him. Meanwhile the wife was in a state of alarm as a flock of birds had settled on her house. An omen of an imminent death? Her husband collapsed in the street on his way home. In this story Jung interpreted the strong psychological state the wife got into on seeing the birds as a synchronicity which he says would be more understandable had she dreamed of the birds. He does not encompass the notion of an omen, which seems, even from a modern perspective, to be the more obvious interpretation. However, as I have already noted, Jung was working in a causally-oriented world although he recognized the acausal. The concept of omens, which today seems ridiculous and is only encountered in the worst sorts of horror film, would have been too occultist for Jung to have mentioned in his essay on synchronicity. This view may perhaps have been strengthened by his close association with the physicist Pauli over acausality, although interestingly enough it is frequently said that Pauli himself was always 'causing' laboratory mishaps and experimental failures.

In *A Mantis Carol* there is mention of an omen. Van der Post had reached Houston, Texas on his visit to the US, a journey initiated by a dream of a mantis, and was staying in a house outside the city. On returning to the house his hostess saw something on the doorstep. It was a preying mantis. None had been seen there before. Van der Post knew what it was. Even more he knew what it meant. 'Even if I had any doubts left about the wisdom of undertaking the journey I think that this appearance of mantis in person on the scene might fully have removed them ... I was travelling as it were under his auspices.'

The power of an omen comes from the symbolism imbued in the external event which triggers the coincidental response. The significance of a synchronicity also comes from its symbolism, as does the meaning of a coincidence (although rather more weakly for this latter case). It could be argued that moving from coincidence to synchronicity to omen involves a concentration of the power of the symbolism inherent in the circumstances of the event and, because the acausal meaning of symbolism is difficult to comprehend, the easier it is to accept the least symbolic of these experiences and the harder to accept omens. Indeed I have already said that many scientists would prefer 'other causes' to explain coincidences rather than to admit an acausal view of the world, so how much harder would it be to admit omens as a part of reality? What I want to argue now is that coincidences, and other manifestations of the complexity of time, are more readily explicable in a symbolic rather than in a materialistic* description of reality. Indeed I would add that coincidences only make sense if a symbolic attitude toward nature is adopted rather than a materialistic one which is also to say that phenomena should be approached through a descriptive rather than an instructional science. But what is meant by a symbolic view of reality? Firstly, such a view implies a hierarchical structure to reality, through the living, the conscious, to a spiritual reality 'above' the material world. It also implies that things lower down the scale are reflections of things higher up.

---

* I am contrasting a symbolic world view to a materialistic one, using the word 'materialistic' in the sense of perceiving physical reality as all of reality and not necessarily in an economic or moral sense. I am suggesting there are two opposite world views or cosmologies, the materialistic and the symbolic.

Consider, for example, the game of chess. This is not just a game of complicated strategies but can be seen as a ritual laden game of life. It can symbolize the conflict between darkness and light, angels and demons, a struggle for the domination of the world. The chequered board is a manifestation of the criss-crossed pattern of life, alternating between good and bad, fortune and ill-fortune. The pieces symbolize aspects of the forces of nature. Not the forces found by science but the symbolic forces of spirituality (the bishop), temporal power (the castle), and so on. The moves themselves are not arbitrary. The knight, who represents the initiate, the force of intellect, moves by jumps of intuition, combining the direct with the diagonal, the rational with the non-rational movement. Pawns are ordinary men, whose path is slow and restricted, whose goal is the state of enlightenment, symbolized by the Queen herself. On the symbolic board, whose eight by eight measure is itself symbolic, all possible moves can be made, all of reality ritualized there, restricted physically to a small field of action. Even time is symbolized by each changing move.

As with chess, so with all the physical environment. Nature seen as a book contains meaning which lies over and above the significance of the individual words and letters written in it. If the words and letters are the equivalent of physical reality then the meaning of the sentences, paragraphs and chapters are the equivalent to the meanings contained in the symbols seen as physical reality.

Viewed in a symbolic light, omens are no longer strange or ridiculous. If all of physical reality has a symbolic meaning, if all material things mirror things spiritual, then omens are much like everything else. The action of an omen brings its symbolic message to the person who encounters it by striking a resonating chord within that person's psyche. The significance of an omen is the coincidence in time between its message and the receptivity of its receiver.

Synchronicities also make sense in a symbolic world. The coming together in time, the contrivance of nature to produce an effect, the acausality, make no sense in a materialistic world view, because such a view precludes nature having purpose. To adopt a symbolic view of nature, to recognize synchronicities and omens, does admit to spiritual reality but it also makes sense of experiences that obviously do have meaning but which our cultural perspective sometimes cannot allow.

Other coincidences also make more sense in a symbolic representation of nature. Time itself symbolizes an aspect of a deeper reality and the coming together of the unexpected and meaningful (meaningful because of its symbolic depth) in time may just be nature pointing out time's own meaning. In Eastern traditions 'Time has engendered everything that has been and will be' and in its progression 'time destroys the world'.

So the coming together of significant events in time, the participation of people with the quality of a particular moment in time ties time to man in a special relationship. Coincidences enable a new perspective on time to be perceived and a new perspective on the world revealed.

8

THE EXPERIENCE OF
TIME

The two ways of describing time – as a quantity and as a quality – can be compared to spaghetti. A quantitative approach is like uncooked spaghetti, hard and inflexible. When one end of a stick of uncooked spaghetti is pushed, something can be manipulated at the other end; there is a positivity about it. Cooked spaghetti, on the other hand, is more like quality. It is soft and flexible as well as more edible. However, when cooked spaghetti is pushed the only thing that happens is that the material gives way; no positive action ensues.

Quality and quantity are somewhat like the ingredients of descriptive and instructional science, respectively. Because the two approaches are so different the sorts of evidence employed in each will also differ. Whereas the instructional approach requires, indeed demands, rigorous, quantitative and reproducable evidence, the descriptive attitude, which often deals with the unique and individual, is mainly anecdotal. This does not mean it is uncritical or sloppy, but in trying to find the whole truth everything must be taken into account. If some evidence turns out to be false, that too is part of the picture. In instructional science anecdotal evidence, *even if true*, can be dismissed as unquantifiable and impossible to assess.

The techniques of instructional science cannot handle individual experience or admit to the quality of time. Descriptive science can; so in this search for time, in its complexity, I turn to anecdotes, which come either from my own experience of time or from what must be regarded as reliable sources. They can be judged not on a quantitative basis but on the fact that they resonate with our own personal experience of time. The fact that the experience of time is not quantifiable puts it into that arena of human perceptions that are at once richer and more meaningful than are those things that are merely quantifiable. As René Guenon has shown, that which can be described in terms of quality is something containing Essence, something approaching Divine truth, while that which is only quantifiable contains purely substance. The lack of quantification of temporal experiences is not something that should stand them in low stead, to be dismissed as nothing more than fleeting perceptions or as merely anecdotal; rather that lack should be seen as their strength. It is because the experience of time is not quantifiable and not subject to numerical comparison that makes it something of quality, something containing the essence of being. It is in search of that quality that this journey continues.

One area in which time's quality may be sought is in considering the 'feel' of different times and ages. Early morning has its distinct quality. Those who rise before most people are up, in any city, find the awakening day has its special character of freshness and expectation. It is quite different from, say, the afternoon, when a certain heaviness lingers in the air. Then compare a Sunday morning, when there is a stillness about, a reverence in the quality of that time, with a Friday afternoon when people are busily preparing to finish work early to start the weekend. If one was suddenly transported into such a time its quality would be clearly recognized.

Then the seasons bring their own special characteristics. Autumn evenings change in tone from September through October to those misty, damp, atmosphere laden November nights. Even then there is still a softness in the air distinct from the crisper tone of December. Of course not all places have such clearly defined seasons as does England but nevertheless the qualities of different times exist. Such qualities are not absolute, for here is another relativity of time. Looking at old photographs, those timeless encapsulations of time laden memories, reveals the quality of the times. It is not just the fashions, though these reflect their ages, nor is it just the appearances of streets, houses and landscapes, but something else comes through which can only be described as the feel of the time. Those 1940s movies could only come from their time and their modern remakes can never quite recapture the right feel of that era.

Then there are times when something seems to be 'in the air', some quality of the time that infects people working at the same time in different fields, or in different countries, but who come up with equivalent ideas and concepts. Newton and Leibnitz invented the calculus at the same time because the time was right for such an idea. In the sciences in the late nineteenth century, for example, one of the main focusses of attention was the phenomenon of light, its colours and its interactions with objects. In the arts the Impressionists were painting light, loosening form to explore how it dazzles the eye in its multitude of hues and colours. With Cezanne, followed by the cubists, painters turned away from the surface interactions of light to examine fundamental and underlying structure and how forms change with time and perspective. In the sciences the underlying structure of the atom was explored and relativity was introduced. At the same time Stravinsky was breaking open the structure of formal musical composition. In the 1970s physicists were conceptualizing about virtual particles in a description of reality where only the ephemeral exists; to counterpoint this movement was conceptual and minimal art. Even the Big Bang theory of the '60s had its twin in 'pop' art!

It is very difficult to see how or why these similarities and comparabilities come about. The links between the groups who come up with the same thing at the same time are usually at best very nebulous and yet the discoveries or creations themselves do reflect the quality of their times. It is something recognizable which can be partially described but for which there is no explanation, at least not in the sense of being able to explain how the clock paradox comes about in Special Relativity, for example. This elusiveness itself is part of the experience of time.

Just as different times and ages have their own quality so also do different places. Just as in considering the time scale of the universe I mentioned the record of times past in the geological strata, which form a sort of physical memory of long ago events and conditions, so too in discussing time's quality an analogous temporal memory can be found in old buildings, unaltered landscapes and prehistoric sites. Into such places a composite of the quality of the different times that were involved in the active history becomes embedded. Age itself is not everything that contributes to the feel of the time about a place. The skyscrapers of New York are imbued with the essence of their time. New buildings may not have a history but the land on which they are built does and the good architect can reflect that quality of time and place in his construction.

There is no obvious structure to these facets of time's complexity. In contrast stands the concept of linear time, nicely quantifiable, divided up into its succession of days, minutes, seconds and even micro-seconds. But the linear view of time does not explain, nor even allow for a description of the quality of time. One attempt to find a quantitative description at least of biological time (if not the total experience of time) which has a qualitative element to it as well comes from the work of Rodney Collin. Collin compared the rate at which time is perceived in early childhood with that in old age and saw that time seems to flow much faster later in life. Biologically internal clocks and metabolic rates slow down as life progresses. Changes in the physical bodies and in the building up of memories and experience are much more rapid in early childhood and youth, and are at their fastest between conception and birth.

Collin recognized that conception establishes itself in the period of one lunar month, the menstrual time cycle, and that life continues for about one thousand lunar months (about seventy seven years). He therefore devised a logarithmic time scale for the human life starting at one lunar month and extending to one thousand. Logarithmic scales put, as equal intervals along the scale, quantities of the powers of ten. Hence the linear scale 0, 1, 2, 3 becomes the logarithmic scale 1, 10, 100, 1000, as shown below.

0	1	2	3
1	10	100	1000

Using this scale Collin realized that the two intermediate points coincided first with birth at around ten lunar months and the peak of childhood at about one hundred months or seven years. So a lifetime on the logarithmic scale is made up of three equal parts: gestation, childhood, and maturity.

1	10	100	1000	lunar months
Conception	Birth	7 years	77 years death	

Gestation   Childhood   Maturity

The value of this logarithmic scale for quantifying a lifetime is that the three periods of equal interval coincide pretty well with equal intervals of temporal experience, in that time seems to pass about ten times more slowly for a six year old child as for a sixty year old man. The scale accommodates a useful representation of the outline of the experience of time and hence a modicum of time's quality. Its inexactness matters little for its concept as a model relies on the way it scales time and not in its precision. To that end a logarithmic scale of time, if a scale is going to be used at all, at least coincides with metabolic rate and therefore to some extent with human experience.

However, man's experience of time is not the same as biological time, so even Collin's logarithmic scale, although an interesting sidelight on this discussion, has little of substance to offer in an attempt at describing the variety of time's multifaceted phenomena. It scales experience, for all its limitations, very much like York time scales duration in cosmology.

Transcending the yoke of time, that burden man has placed on himself with the invention of clock time, is a valued human experience. To transcend time is also the aim of meditation and other spiritual exercises. Krishnamurti wrote:

To meditate is to transcend time. Time is the distance that thought travels in its achievements. The travelling is always along the old path covered over with a new coating, new sights, but always the same road, leading nowhere – except to pain and sorrow.

It is only when the mind transcends time that truth ceases to be an abstraction. Then bliss is not an idea derived from pleasure but an actuality that is not verbal.

The emptying of the mind of time is the silence of truth, and the seeing of this is the doing: so there is no division between the seeing and the doing.

Krishnamurti's time as the distance thought travels parallels the idea of time as the distance light travels. The emptying of the mind of time in meditation is a way of controlling the unexpected and often awesome experience of timelessness, but done in a deliberate way and toward a particular end. This suggests that many temporal experiences are part of the same spectrum and produce an awareness of time as the limitation of consciousness. It is only when it is recognized that the temporal experiences which stand out in one's awareness are pointing to something that they have a meaning beyond their appearance, that time and its many faces can be put into a truer perspective. The manifestation of time is not random, it everywhere seems to display a purpose.

In searching for the quality of time several diverse areas of experience have been examined in order to obtain a 'feel' of that quality. I want to turn now to four areas in which different faces of time are displayed and which, although qualitative, reflect aspects of time already encountered in earlier chapters. The first of these has already crept into this encounter and turns out to be a more common experience than might be expected. It is the experience of timelessness, of time standing still, of someone stepping outside time. In the fascinating collection of material at the Religious Experience Research Group at Manchester College, Oxford, are several accounts of such experiences, some of which were selected for the anthology on religious and mystical writings by Cohen and Phipps, and entitled *The Common Experience*. I have drawn from that source for this account of the experience of timelessness.

I was in the garden, muddling about alone. A cuckoo flew over, calling. Suddenly, I experienced a sensation I can only describe as an effect that

might follow the rotating of a kaleidoscope. It was a feeling of timelessness, not only that time stood still, that duration had ceased, but that I was
myself outside time altogether. Somehow I knew that I was part of
eternity. And there was also a feeling of spacelessness. I lost all awareness
of my surroundings. With this detachment I felt the intensest joy I had ever
known, and yet with so great a longing – for what I did not know – that it
was scarcely distinguishable from suffering.

How long I stood, or would have gone on standing, I do not know; the
tea-bell rang, shattering the extra dimension into which I had seemed to be
caught up. I returned to earth and went obediently in, speaking to no one
of these things.

The timeless moment, which takes a person out of time in the
worldly sense, is often closely associated with religious or mystical
experience and altered states of consciousness which may also be
induced by hallucinogenic drugs or high fever. And yet it is a very
common experience.

Timelessness can take the form of time and motion ceasing. The
stillness that enfolds Sleeping Beauty and all the people and
creatures in her castle for a hundred years is a version of the timeless
experience. Another is found in the Apocryphal Gospel of James:

Now I Joseph was walking and I walked not. And I looked up to the air and
saw the air in amazement. And I looked up unto the pole of the heaven and
saw it standing still, and the fowls of the heaven without motion. And
behold there were sheep being driven and they went not forward but stood
still; and the shepherd lifted his hand to smite them with his staff and his
hand remained up. And I looked upon the stream of the river and saw the
mouths of the kids upon the water and they drank not. And of a sudden all
things moved onwards in their course.

Memories often contain an element of timelessness and so can
works of art or objects from a past age, which themselves are
timeless in the other sense, transcending time corporeally. The
timeless experience is a real connection between time bound
mortality and the timelessness of eternity, beyond and outside time.

Eternity is often thought of as being a very long time indeed, an
infinity of time, but St. Thomas Aquinas pointed out that eternity
was outside time completely. Eternity means lack of succession,
lack of change, an instantaneous whole. God, he wrote, is eternal
while hell is unending. Aquinas' cosmology was earth centred, with
the great spheres of the planets surrounding the earth, and it lasted
until Copernicus. The outer sphere governed time in the lower
regions, including earth, and beyond that sphere was eternity.
Between the two, the time laden earth and timeless eternity, came

the time of the angels, time bound but imperishable. Angelic time, as described by medieval writers, is like a ray of light which has duration but not sequence. Henry Vaughan, the seventeenth century poet, also connected light, time and eternity in his poem, 'The World':

> I saw eternity the other night
> Like a great *Ring* of pure and endless light,
>   All calm, as it was bright,
> And round beneath it, Time in hours, days, years
>   Driv'n by the spheres
> Like a vast shadow mov'd, in which the world
>   And all her train were hurl'd.

The connection of light with timelessness has already been examined in a rather different context, and yet there are distinct similarities between the timeless nature of light and the experience of timelessness. To the observer, light travels through time and ordinary time ticks on. Yet to light and the person experiencing timelessness, time stops still, motion may cease, and a link is forged with the eternal. Some cosmologies and some scales of time can cope with and encompass the infinite and endless successions of time but none include the eternal, for not even outside the light cone does time itself cease. It is only through light that time ends.

In contrast to the timeless experience is that of deja-vu, of time repeating itself or at least its memory recurring. The first example of deja-vu that I can remember now, although I do not believe it was the first to happen to me, was when I was about twelve years old. I remember coming in the back door of our house and calling out to my mother, who was upstairs, that I was back, and was then overwhelmed with the feeling that this moment had happened before. I knew instantly that my mother would call out that we had got 'salad for dinner and then, of course, she did so. Such an experience lasts usually for a few moments, but the intensity can be dramatic.

A few years ago I was teaching a student physics in an upstairs lecture room rather than in my office, which I shared with someone else who wished to remain there. I had reached a part of the tutorial where we were discussing radioactive half lives and I was again swamped with the deja-vu feeling. I knew I was going to suggest that I needed to show him some examples from a certain book in my office and then go to collect it. I resisted saying this to him, but the feeling it had all happened before was strong. I was determined to break the pattern of the event. I turned to my student and asked him if we had done this work before believing that he might be sharing the experience. He looked puzzled and replied no. I struggled to avoid continuing the experience. I resolved not to go and fetch the book, that I knew I had done the 'previous' time. Having made that resolution I turned again to the student and said: 'I think I had better show you some examples of these. I will just pop down to my room and get a book.'! My awareness of the experience itself did not make it go away, even when I tried not to repeat its pre-set pattern. There is an element of precognition itself in the experience, because the situation is so 'familiar' that one knows what will happen next. It is different from precognition, however, in that it is familiar; one is in a sense reliving a part of one's life, not predicting or sensing a remote event.

There are many explanations for this sort of temporal experience involving either altered states of consciousness or parallel universes. Others contend that all times are eternally present and that in the experience one becomes momentarily aware of this. Maybe the mechanism in the conscious mind that inhibits an awareness of eternity (or at least of all other times) as existing now becomes switched off or operates incorrectly to give one the experience. However, I do not know that I like these explanations. It seems to me that deja-vu, like the experience of timelessness, is a manifestation of the multifaceted nature of time which is perceived in a non-linear, maybe subconscious, fashion which intrudes into consciousness. Just as people can go outside time, so too can they occasionally experience time indirectly, sideways in a sense, and hence pick up something else of its quality.

On a space/time diagram no two things can ever be exactly alike, no two events can ever be exactly at the same place in space at the same time. Time always moves on, so that even lack of movement in space still involves movement through time. In the experience of deja-vu, however, time repeats itself. There may not be similar causes but similar effects occur. Somehow the experience moves the person to whom it occurs askew and carries him through the same again.

This motion through time also manifests itself in the experience of time slip, of passing into another time. Unlike deja-vu where the other time was familiar, part of one's own past (or future) experience, time slip involves an unknown experience in someone else's time.

The classic example of a time slip involved two senior Oxford academic ladies, Charlotte Moberly and Eleanor Jourdain, who in 1901 were visiting Versailles Palace. Both women experienced a depressing and dream-like state without realizing anything was unusual, and yet they found themselves in the Versailles of 1789. So natural was the experience for each of them that it was only when they later discussed the visit that anything odd about it emerged. Somehow they got muddled up in the wrong time.

Colin Wilson tells of a similar case involving a lady named Jane O'Neill. Miss O'Neill suffered an accident and subsequently had visions and a clairvoyant ability. When visiting Fotheringay Church in east Northamptonshire, she was impressed by the painting behind the altar. Her companion later remarked that she could not remember the picture. A phone call to the parish confirmed that no painting was there and the mystery deepened when Jane O'Neill revisited the church. The picture was not there, and the church itself seemed distinctly smaller. Only when a local historian was consulted did she discover that she had seen the church as it was in the early sixteenth century, before it had been rebuilt in 1553. Like Miss Moberly and Miss Jourdain, Jane O'Neill had experienced a displacement in time and had temporarily lived in another era.

Two things are interesting about these cases. First, the people concerned had no direct sense that anything unusual had occurred to them. They seem to have slipped into another time without noticing. The second point is that in both cases, and in other similar anecdotes, the time slip is always into the past. Several questions come to my mind concerning this experience. How common is time slip? Could it be that the deja-vu experience is connected with the phenomenon? It may be that the events of a deja-vu are remembered because that piece of time has been slipped into before. This would enable a time symmetry for these displacements to be displayed. The forward time slips are not remembered because memory is trained to remember the past, but they could occur.

The other intriguing question is this. If the ladies found themselves in the park at Versailles in 1789 were they visible to the people who were 'normally' there at that time? If so, how did they look? Were they dressed in clothes of the future? Did they appear and disappear? To the ladies of 1901 the experience was one they lived through, but if they stepped into another time they either did so as a sort of ghost or else as real live people. As the idea of ghosts is more familiar than the notion that people we see may not really be here and now all the time, but just visiting from the future, the ghost idea is perhaps more acceptable. For, leaving aside the spiritualistic aspect of ghosts and hauntings, ghosts are in a clear sense caught in some form of time slip. When a ghost is seen, it is seen as an apparition of someone as he was in another time. Communication with such apparitions raises some problems, but if the event is considered as a time slip the ghost is here and now although displaced from his proper time.

This discussion is as haunting as that encountered with virtual particles and the time reversal of the sub-atomic world. Yet the time slips of the ladies at Versailles were macroscopic events and not illusions of a virtual reality. It is inconceivable to imagine that at that time and place sufficient sub-atomic events all coincided to reverse the time for two ladies in such a way that their personal time was transported. Despite several attempts by investigators to link quantum uncertainties with paranormal activities, two of which may be ghosts and time slips, the attempts have not succeeded. It is not surprising because the theory is a mathematical, quantitative description and the experiences are reflections of the quality of time (and the quality of space, matter, and so on).

A still more complex manifestation of time is told by Matthew Manning in his book, *The Strangers*. In 1971 the young psychic first met Robert Webbe. Webbe had built half of the house the Manning family had moved into near Cambridge. The house itself was part seventeenth, part eighteenth century and Robert Webbe was a ghost. He first appeared to Matthew as a person, quite solid and real, on the staircase. He had bad legs and walked with two sticks. He said: 'I must offer you my most humble apology for giving you so much fright, but I must walk for my blessed legs.'

As a young teenager, Manning had been the focus of dramatic poltergeist activity. One way he found of controlling that wild 'force' was through automatic drawing and writing. In this he would sit with pen and paper and be 'guided' in drawing or writing. It seemed natural for him, therefore, to try to communicate with Robert Webbe via automatic writing, and this he did. The conversation quoted later was written in this way.

Webbe was seen quite frequently, communicated with regularly and his presence was made known in several other ways. He lit the Manning household candles, he gave them presents, material objects which he left for them (technically known as apports). A loaf of bread appeared, a tallow candle, pages from books, trinkets. He 'stole' some pictures from the wall and hid them in his 'safe', a compartment under the bedroom floor, unknown to the Mannings until Webbe told Matthew where it was. The floor board was screwed down on top of the compartment but inside were the pictures.

In May 1977 Manning had his last conversation with the ghost. During this talk Webbe was asked if there were ghosts in his house.

Of course not else I would chase them away, he quickly replied.
We live in your house now and we have seen a ghost there.
I do not believe such tales. You try to frighten me out.

Then Manning asked the crucial question.

Who do you think you are talking to?
I think sometimes I am going mad. I hear a voice in myne head which I hear talking to me and asking me what I do. But tell no one else they locke me away. Who is this voice?
This voice is me, who am I?
You frighten me. Who are you? I only hear you in myne head and not in myne ears. Who are you? Are you the ghost this voyce talks of?

Manning then told him that it was he who was the ghost.

You are mistaken, I am no ghoste. I am here. You frighten me. And who do you say you are? ... Are you a ghoule of tomorrow?

This extraordinary encounter between Manning and Robert Webbe is more than just another anecdote of a haunting. It has far more of a time slip feel to it. But the question that needs answering is who is slipping into whose time? Certainly the physical manifestations, the apports, must be slips by Webbe into the 1970s, but the conversations are just as much a time slip by Manning into the 1720s. Webbe in 1726 was hearing voices in his head. Manning in 1977 was writing down Webbe's thoughts. Time became dislocated, merged. The fact that Webbe could not see Manning suggests that ghosts from the future are not visible, perhaps because at the earlier time they have not yet acquired a physical form, and that ghosts from the past can bring their earlier appearance with them. In that case the ladies at Versailles would not have been seen. However, the implication that ghosts are not just things of the past is as queer an idea as any. For me, this story is perhaps the most telling of all the anecdotes concerning disrupted time I have come across; its implications are remarkable. I will not suggest 'mechanisms' to explain this disruption in time but again point out that what is being witnessed is the complexity of time. Time is being interpenetrated and is interpenetrating through the fabric of 'normal' reality. The peculiarity of this interpenetration exceeds that of, say, gravity, whose possible action at a distance effect could be explained by advanced waves moving back in time in counterpoint to their progressive and retarded partners. This interpenetration slips forward and backward in time simultaneously, seeming to defy the laws of matter and causality. Yet time and time's quality are closely tied to the acausal side of nature, where defiance of causality is commonplace. Pursuing the quality of time is as strange as following where its quantitative description led.

A haunting similarity in past and future, focussed on two people's 'now,' reflects an idea Laurens van der Post wrote of in connection with the mantis. Whenever that insect deity is in trouble it is said he dreams a dream to see him through. A Bushman hunter then told van der Post, 'You see there is a dream dreaming us.'

9

SEEING THROUGH TIME

To slip sideways through time is one thing, to see through time into the future is quite another. Time yet to come appears to be far more opaque than does time past; and yet in a variety of cases time seems to become transparent when certain people see through it to events that have not yet occurred. This aspect of time's complexity, this transparency that is revealed, demonstrates another side to time's paradoxical nature. The experience of seeing through time also seems a common one and its many manifestations will be examined in this chapter. Through such varied phenomena as precognition and prophecy, prediction and divination, time displays some of its most curious effects.

Consider the dream J.W. Dunne had in 1902, when he was serving in the infantry in South Africa.

I seemed to be standing on high ground. Here and there in this were little fissures, and from these jets of vapour were spouting upward. In my dream I recognised the place as an island of which I had dreamed before – an island which was in imminent peril from a volcano. And, when I saw the vapour spouting from the ground, I gasped: 'It's the island! Good Lord, the whole thing is about to blow up!' . . . All through the dream the *number* of the people in danger obsessed my mind. I repeated it to everyone I met, and, at the moment of waking, I was shouting 'Listen! Four thousand people will be killed unless. . . . '

I am not certain when we received our next batch of papers but when they did come the Daily Telegraph was amongst them, and, on opening the centre sheet, this is what met my eyes: VOLCANO DISASTER IN MARTINIQUE. PROBABLE LOSS OF OVER 40,000 LIVES.

Curiously, Dunne misread 40,000 for 4000 on learning of the Martinique disaster and did not realise his mistake for several years. This error had also been made subconsciously and had led him to foresee the number incorrectly in his precognitive dream. The whole incident is, of course, well known as it was on the strength of this and similar incidents that led Dunne to investigate whether precognitive dreams were common. When he found other people responding to his accounts of precognition by recalling the cases they too had experienced, he set up his *Experiment in Time* to check how widespread precognition really was. He wrote down his dreams systematically and asked others to do so too. The written dreams were later compared to the events they had foretold by checking back over the recorded dreams every few days or so. The degree of success was variable, but it emerged that a fraction of all remembered dreams seem precognitive, though frequently in fairly trivial ways. One lovely example, combining both the 'triviality' of the event with Dunne's requirement for precise details, involved a dream by his cousin. She dreamed of a German woman with her hair in a bun, dressed in a black skirt and a black and white striped blouse. In the dream the cousin met this lady in a public garden and thought she was a spy. Two days later the cousin went to a country hotel and was told of a 'curious person staying there whom all the guests suspected', a German woman. The cousin later met this lady, in the rather formal gardens. She was dressed in a black skirt, black and white striped blouse and her hair was done up in a bun!

There are two aspects of precognitive dreams that I would like to mention, the first of which involves the extent of the detail presented in the dream in relation to the event foretold. The truly precognitive dream must contain individual details which are sufficiently close to those of the event. Dunne realized this with his own experiences of time and stressed that it was expressly in the specific details that the precognitive was to be recognized. Dreaming about finding some money on a wall will not be sufficiently detailed to count as precognitive when the dreamer finds some money on a wall two days later. But if the dream specified the number and type of coins and placed the wall in a specific context, such as being beside a wrought iron gate and next to a broken wooden shed, and the experience was exactly that, then the dream would have been unquestionably precognitive.

The second aspect of this phenomenon involves memory, which can be surprisingly deceptive. It is seldom sufficient evidence for someone to rely on the memory of a dream when faced with an event they remember dreaming about. It is all too easy to remember a dream in the light of the later experience and, except in those rare cases where the dream itself was exceptionally powerful and impressed itself firmly in the mind, it is wiser to be cautious in relying on a reminiscence.

One sceptic in these matters was Rudyard Kipling, yet he recounted how he had a dream of himself standing alongside other men, all formally dressed and lining a hall where a ceremony was being conducted. At the end of this ceremony Kipling dreamed that a stranger approached him, took him by the arm, and said that he would like a word with him. Some weeks later the author was at a memorial service in Westminster Abbey when he realized he was living out that dream. The details were exactly as foreseen. Most remarkable of all was the fact that when the service was ended a stranger did come up to him, took him by the arm and said, 'I want a word with you, please'. Kipling described the event in terms of his being shown 'an unreleased roll of my life film'.

What this and all other precognitions question concerns the nature of the future. Does it lie ahead of 'now' like film already exposed, which can be surreptitiously glanced at in advance, or does the future not exist 'now'? If the latter is the case, then surely dreaming about an event that does not exist and then living through that event means the two occasions are only connected by chance: that the event was the same as the dream by coincidence.

The previous discussion about chance and coincidence would suggest that this case is not likely, that seeing through time is not just chance speculation but is alternatively a real manifestation of one of time's qualities. The connection between the present and the future is however, acausal. For it to be causally connected would violate the restriction placed on the sequence of events that is limited by the speed of light inside the light cone. To understand precognition at all must be to switch to an acausal reference frame in which events can be juxtaposed without a cause.

Another question which arises is whether the knowledge gained by seeing the future event could alter or even avert the event. If it can, then the vision is hardly a true precognition for the event foreseen paradoxically never occurs. However, this question does raise immense problems, as the following account shows.

While visiting the US, Matthew Manning had a strong precognitive dream which involved a plane crash near an airport. In the dream the plane was flying too low and had to rise to clear some trees. As it did so the right wing dipped and smashed into a pylon. In the dream the wreckage was seen scattered over the ground near the runway and across a marsh. The precognition was both powerful and shocking as the foreseen disaster involved much loss of life.

Manning told his dream in detail to his companion, even drawing a sketch map of the crash location and these details were mailed back to England. An agonising problem then had to be faced; armed with the details of this potential air crash what could Manning do? He would first have to locate the airport near which this crash was purported to occur. It could be anywhere. But assuming he could find out the location, what then? He did not know when the disaster would strike, to which aircraft, or any such detail. All he knew was how the crash would occur and its anonymous location. Even if he had known the exact flight, what could be done? He could have told the airline and they could have cancelled the flight, or re-routed it or instructed the pilot to be extra careful. It is difficult to know whether anything could have been done that would actually have enabled the event which was 'seen' to be prevented. However, a well known case where forewarned was forearmed involved a motor car accident in which the dreamer was driving down a particular road under specified circumstances when a child ran out in front of the car and was knocked down and killed. The reality of the dream materialized shortly afterwards, when the man found himself driving down a road under the same circumstances. The child did run out into his path, but the driver had already recognized the situation and was even then applying his brake so the child was not harmed.

This incident certainly implies that precognition does not necessitate loss of will to alter a possible future event. It does pre-empt the notion that the future is entirely predetermined, like an already exposed film, at least to the extent that precognition of the future does enable its alteration, if circumstances so permit. That does not mean that Dunne could have prevented the volcano from erupting, for that was a precognition of an event he was not directly involved in. The same was true of Matthew Manning's dream about the aircraft accident.

That crash did in fact occur four days later. A Boeing 747 landing at Kennedy Airport swung round and dipped to the right, lost height and struck an electricity pylon, crashing into marshy land exactly as Manning had plotted in his sketch map. The crew and passengers, 122 people, were all killed. John Taylor, in his book *Science and the Supernatural*, recounts the precognitive dream of a woman. It was a nightmare in which she was trapped in a tube, poorly lit and filled with smoke and screaming and crying people. The dream awoke the woman, who told her husband of her nightmare. The following day the worst disaster in the history of London's underground railway occurred, the Moorgate disaster, when a train had smashed its way into the end of a tunnel siding. It seems as if highly charged emotional events trigger off precognitions. The question then arises as to whether events somehow 'radiate' outwards in time, after the event as memory and before the event as precognition. People then seem to 'pick up' this reversed memory of the event before it happens. As Matthew Manning says: 'The anguish of a disaster can be "audible" before it occurs.'

Certainly the most dramatic precognitive dreams are those involving human disaster, and the strength of the precognition seems related in some way to the horror of the disaster. But nothing clearly emerges in trying to analyse precognitive experiences. The examples looked at here have preceded the events by a matter of, at most, only a few days. In Dunne's experiment he regarded two days as an optimum time lag between dream and event, unless the precognition was very strong. But in his case a precognition of the First World War bombardment of Lowestoft by the German navy came a year before the event. It seems, therefore, that there is no clear correlation between strength of precognition and the dramatic content of the foreseen event or one that connects the time delay between precognition and occurrence. This is not surprising because such a correlation would only be anticipated if a causal principle was in operation. In precognition the effect, or at least the result of an effect, is known before the cause. Causality cannot be at work, at least as far as causality is understood.

I now turn to another example of the way time plays tricks, an anecdote told to Arthur Koestler by Sir Alec Guiness and it involves not just another form of precognition but a fascinating view of the experience of 'clock time'.

Saturday July 3rd 1971 was, for me, a quiet day of rehearsals ending with dinner with a friend and going to bed at 11.30 p.m. Before going to bed I set my two alarm clocks to wake me at 7.20 a.m. When working in London at a weekend it has been my habit to get up at 7.20 on the Sunday morning and leave my flat at 7.45 for the short walk to Westminster Cathedral for Mass at 8.00. (I have been a Catholic, of a sort, for about sixteen years.) On returning from Mass I would have a quick light breakfast and catch the 9.50 Portsmouth train, from Waterloo, to my home near Petersfield. On this particular night I remember I didn't sleep a great deal as I constantly woke up – perhaps each hour – with a tremendous sense of well-being and happiness, for no reason that I can put my finger on.

By habit and instinct I am a very punctual riser in the morning, and usually wake up two or three minutes before the alarm clock rings. On this particular morning I woke, glanced in the half light at the clock and thought 'My God, I've overslept!'. It appeared to me the clock said 7.40 (I didn't refer to the second clock). I rushed through washing and so on and hurried to the Cathedral. Very unexpectedly – in fact it had never happened before – I found a taxi at that early hour, so I thought I was at the Cathedral at 7.55. With time to spare I went to confession. When Mass started I thought the attendance was considerably larger than usual for 8.00. It was only when what was obviously going to be a rather tedious sermon was under way that I glanced at my watch and realised I was at the 9.00 Mass instead of the 8.00. I went home, as usual, saw that both my alarm clocks were correct and decided to catch the 10.50 train instead of the 9.50. (My wife was away in Ireland so it made no difference what train I caught.) When I arrived at Waterloo at 10.30 there was an announcement that all trains on the Portsmouth line were delayed for an unspecified amount of time. An enquiry gave me the information that the 9.50 train had been derailed a few miles outside London. Subsequently I found out that it was the front coach of the train which had toppled on its side and that, although no one was killed, or even grievously injured, the occupants of that coach had been badly bruised and taken to hospital. My habit, when catching the 9.50 on a Sunday morning, had been to sit in the front compartment of the front coach because, when in Waterloo station, that coach was in the open air, away from the roofing of Waterloo and consequently with more light for reading and less likelihood of being crowded. . . .

The remarkable thing about this story is not just the avoidance of the accident, but the misreading of two adjacent clocks. The story is full of references to specific times and, although Guiness realized he had overslept by twenty minutes (in fact sleeping through two alarm clocks going off), he in practice had overslept by rather more than an hour. It is interesting, too, that his habit was to wake two minutes before the alarms normally went off, another common experience of time, in which the unconscious guides consciousness through an indirect sense of time. I am reminded here of Salvador Dali's curiously draped, soft clocks in his famous surrealist painting, visually displaying how the unconscious views time quite differently in quality from normal, rational thinking.

At this point I had better draw some distinctions between three different ways of seeing through time. I have already discussed and will return to the subject of *precognition*, the ability of perceiving a future event. Quite separate and distinct from that is *prediction*, which is the ability to speculate about the future. *Premonition* is yet another way of foretelling which involves an emotional response to a future event, a hunch that maybe something is wrong, although it seems to involve neither the detailed knowledge that comes with precognition nor the sight of the future event. Premonition is a special kind of precognitive experience, an inexact foreboding. Sir Alec Guiness sleeping abnormally through two alarm clocks was a form of premonition. After disasters there are always many cases to be heard of prior fears or odd circumstances that contrived to prevent people from venturing into the disaster area. As the Titanic set sail on its fateful voyage a woman on the quay suddenly shouted out 'that ship is going to sink before it reaches America'. An English newsreel cameraman in the Second World War had a reputation for premonition and 'lucky' escapes. He continually missed or was prevented from reaching a number of ships that were sunk or aircraft that were shot down. His 'sixth sense' served him well.

Prediction, on the other hand, is based either on an educated guess, say about the contents of the next governmental budget speech, or on extrapolation of past occasions into the future. I can predict that the sun will rise tomorrow by extending the run of past cases and applying the known laws of planetary motion. Prediction can, of course, be wrong and frequently is, especially when concerned with human events, but prediction lies at the heart of modern science as well. The proof of science lies in its predictive power. A good scientific theory is expected not just to satisfy and explain known processes, but also to predict novel phenomena that can then be tested. For example, Einstein's General Theory of Relativity predicted that light would be bent by a gravitational body and this was later found to be so. But prediction of this sort applies to general laws and principles and not to the detailed foretelling of a specific, individual event and it is that which is claimed for precognition. Premonition cannot make that claim, for it does not have the same element of detail that is found in, say, a precognitive dream.

Precognitions are frequently and mistakenly referred to as predictions, but can only be thought of as such in the sense that they are predictions based on some precognitive awareness. A number of tests have been made of the predictive claims of psychics. In one of these, for example, a psychic named Orlop was asked to predict who would sit in a particular chair in a hall on a specified future date. A description was given of a lady of a certain age and height, whose occupation was related to entertainment and who had recently injured her leg. When the particular evening was reached and an audience had randomly filled the seats of the hall the prediction was read out and the occupant of the specified chair was asked to describe herself. She was of the age and height predicted, she was an actress and had just injured her ankle.

This sort of prediction, however, is based on the precognitive ability of which psychics seem to be capable almost at will. Others seem to have this gift on odd occasions, especially when asleep. This suggests that consciousness may inhibit seeing the future at the moment of now. Whether being psychic is a basic human characteristic, the sixth sense, is beside the point in this investigation of time. The psychic ability is a method of penetrating temporality in advance, just as memory is a method for reaching to times past. Memory and precognition have similarities, differences and limitations. Although the existence of time past and its recall presents no conceptual problems, the existence of time yet to come and its manifestation is difficult because of time's one way flow, its asymmetry; yet precognition and premonition illustrate that time to come can make itself known, that future times can somehow interpenetrate the present.

This interpretation of times has already been illustrated by the experience of deja-vu, of time slips and mutual haunting. It is worth commenting that the deja-vu effect could be the living out of a precognitive dream, hence the element of knowing what is about to happen next and the total familiarity of the occasion. There seems to be a difference between these two temporal phenomena. In the cases I know the recognition of the precognition is what is experienced. That is, the prior notification of the event is remembered, whereas in deja-vu the experience is one of actually having lived that moment before. Deja-vu seems to be more likely related to time slips, where the temporal displacement is into one's own life rather than into some other piece of the future. The interpenetration of both past and future time is told in Dickens' *A Christmas Carol*. The Ghost of Christmas Yet-To-Come takes Scrooge to his own burial in a vision of what might be; which, like the precognitive, allows the future to be changed in the light of what has been shown of it.

Attempts have been made to investigate precognition in as rigorous and scientific a way as possible. In such investigations the nature of the effect being pursued is usually, from practical necessity, reduced to rather trivial proportions. In testing, for example, a subject's ability at clairvoyance, of seeing what is not physically present, the clairvoyant's task may be reduced to guessing the simple image on one of five possible cards. The laboratory associated predominantly with such card guessing experiments was that set up by J.B. Rhine at Duke University in North Carolina. As well as testing and finding evidence for telepathy and clairvoyance

Rhine and his colleagues also found precognitive abilities in some of their subjects. It was found to be possible to guess the next card in the pack before it was even revealed to the 'transmitter', the person who was looking at the cards and whose thoughts were being 'read' by the subject doing the experiment. The results of these tests have a high statistical significance. Although the precognitive task in these experiments was trivial the effect demonstrated is substantially the same as revealed by the anecdotal examples and by individuals' experience.

Precognition has also been tested by Helmut Schmidt. Schmidt set up an automatic device governed by the radioactive decay of a strontium-90 source to randomly switch on one of four lamps. Subjects had to guess (precognitate) which lamp would be the next to light up. Success in the results of using this apparatus was outstanding, with subjects achieving guess rates that deviated from chance with odds estimated from tens to thousands of millions to one. However, as has been seen, radioactive decay may not be truly random, and the experiment with chicks indicated that the decay rate could be altered by some aspect of 'consciousness'. Caution should be exercised in comparing results with a model of randomness; and there must be some suspicion about the radioactive decay. If it is admitted that the results of the test are still valid and significant, after allowing for the non-randomness factor, then it could be that a psychokinetic effect, mind influencing matter, is being displayed rather than true precognition. Such a result would be valid and important for paranormal research but would not concern precognition. On this level some of Rhine's card guessing tests are more satisfactory in that sequences of cards have been predicted at an appreciable time in advance of the series turning up. Of course it could then be argued that any guess might be precognitive in the sense that the event guessed is bound to turn up somewhere, sometime, but that argument ends up by being merely dismissive and overlooks the qualitative aspect, the degree of meaning inherent in the sorts of phenomena being discussed. Incidentally, the lack of any serious and personal meaning in the sorts of test carried out in parapsychology laboratories, card guessing and so forth, explains, to me at least, the so-called decline effect, where a subject's ability at whatever facet of ESP is being tested for declines with time. It is a sort of psychic boredom. The ultimate reduction of precognitive ability would be to test how successfully a computer could guess a future event. But this test has already been discussed. My random number experiment was, in effect, also a test for precognition. The computer was attempting to guess a random number that had not yet been generated. In practice the results of the tests varied insignificantly whether the computer was trying to guess the following number or the next but one or even one at a variable (random) position later in the generated series. The results of such a test were statistically significant. The computer, too, had precognitive abilities!

Of course this conclusion is nonsensical and the significance of the apparently extraordinary results in random number generation has been discussed. I bring this question into the investigation here only to emphasize the caution with which claims for the scientific 'proof' of paranormal phenomena must be taken. This is not to throw doubt on precognition as such. Rather it is a reminder that the orthodox scientific approach, the instructional science approach, to these problems can not only dilute the phenomena but can also lose sight of what it is investigating under a pile of statistics and in a search for causal connections. I do not doubt that both Schmidt and Rhine have demonstrated precognition but I do repeat my assertion that the penetration of time is essentially related to the acausal world and that a strictly causal or instructional approach to such matters will often confuse the issue, whereas a descriptive, anecdotal approach can clarify.

What is time doing when precognition takes place? This question implies that time is itself an active agent. A passive form of the question would be, instead, to ask how time is manifested in a precognitive act. Does time reach backwards as well as forwards from every event, radiating outwards in both temporal directions like a ripple spreading out across the water of a smooth pond surface? Is precognition the sensing of a 'splash' in time? Certainly Matthew Manning's description of the audibility of human anguish from events concerning great suffering seems to fit into this image for precognition, the image of the 'time splash'. But why should such a 'vibration' emanate from card guessing experiments or other quite trivial events?

I distinguish two sorts of precognition. The first sort, the kind involving dramatic events, on a personal or a collective scale, is a true precognition. It possibly occurs because as well as people's material bodies extending outwards in space they may also extend outwards in time. Let me explain what I mean. Bodies do not stop at the skin but extend outwards into the space around them. Human senses reach out into the world, some farther than others, but even apart from the five senses man extends outwards. People can 'sense' a volume larger than their physical bodies, and whose size depends partly on the person's own emotional state. When feeling happy and expansive I extend out some way, I feel bigger. When I feel insecure I shrink down both psychologically and in the space around me. Everyone feels this protectiveness toward the space around them when bustling through a crowd, partly because that space has to be pulled in closer. Psychics claim to see part of that space as the aura. If people extend in space why not in time as well? What if time was three dimensional, like space, and could extend from now into the past and the future? If people extend outward in time then it would be expected that others could sense them before their arrival and feel their presence diminishing after they had departed. When lives suddenly become involved in extreme drama and emotional tension then the extended selves may well reach out dramatically in space and time, and thereby be 'detected' by people sufficiently aware of their own sensitivity to these things. The operation of such an effect would not be causally connected so 'common sense' notions of causality will not be upset; this is an acausal phenomenon.

True precognition, then, would only operate for humanly significant events. Card guessing has so slight, if any, temporal extension that it is hard to envisage how such events could be precognitively perceived. It could be that all those tests with cards and radioactively controlled light bulbs are only demonstrating something about human gullibility and about the non-randomness of the supposedly random, but if it is conceded that an effect is there, then its explanation more probably lies in a different direction from what I have called the truly precognitive. Rather, I would call this dimension to precognition a precognitive clairvoyance, an ability to see what is elsewhere in advance of its direct perception. Such an experience still involves the extension of man's horizons beyond his physical boundary, but whereas the truly precognitive involves foreseeing others' extension backward in time so does precognitive clairvoyance involve perception by one's own extension forward in time. In one sense the two phenomena are aspects of each other, but I suspect they are sufficiently distinguishable to make this separation in their description.

Another way of attempting to explain precognition could involve the adoption of words like 'radiation' and 'detector' in a more technical sense. Some form of ethereal wave which emanated from events could be postulated, which had both advanced and retarded components. Some human faculty for detecting these waves could then be suggested, which when activated picked up the advanced wave signals and gave rise to precognition. Such an explanation is, of course, reminiscent of action at a distance except that in this case the retarded wave would have to be cancelled and only the advanced part would be active. This is a neat, causal explanation, superficially attractive, but unsatisfactory as it is really nothing more than an *ad hoc* proposition.

Attempting to explain precognition, or any other paranormal event, by means of a causal or instructional theory will inevitably end up being little more than *ad hoc*. It is the ultimate fate of all such explanations that they must end up confronting what in their own reference terms is inexplicable. This is because these explanations do not, by definition, include that part of the natural world that is essentially qualitative and therefore in one way, non-physical. As most interesting phenomena have non-physical components in their make-up, a purely instructional approach will always be inadequate and always fail to fully explain. Time, as a phenomenon with material manifestations (although mostly indirect ones), is a marvellous example of something which is physically inexplicable and which overtly points, through both its connection with light and also its often strange manipulation of physical awareness, to a more complex hierarchy of nature. Recognition of the hierarchical structure of the world is also a recognition of the supernatural, in that the levels of the hierarchy above the highest of the material and conscious planes are literally above nature. The most cogent and readable account I know of about the distinction between nature and supernature is found in C.S. Lewis' book *Miracles*. The penetration of the natural world by the supernatural leads to the recognition and description of events as being supernatural, in that no natural explanation can be found for them.

Precognition and clairvoyance are not supernatural, in that they do not arise from or are manifestations of the world above nature, except in so far as everything below is a manifestation of qualities above in what I have called a symbolic view of reality. Rather these are aspects of the natural world but for which there seems no obvious explanation. Such phenomena can be described as operating alongside normal nature and are called paranormal.

The paranormal is not necessarily supernatural and certainly the range of paranormal phenomena usually discussed does not encompass the supernatural, but if the perception of fairies, angels and divine revelations are included in the paranormal, then these would also be supernatural. A Dominican monk made this distinction in a lecture, which I think is a useful one. It was said that when Saint

Dominic celebrated mass, as he raised the host, the communion bread, he levitated off the ground. The transformation of the bread into the body of Christ, the monk said, was a truly supernatural event; the levitation of St. Dominic was, however, simply preternatural. All preternatural phenomena, he said, were purely illustrative of the supernatural and not supernatural as such. So too, can many paranormal phenomena be regarded as illustrative of the complex and hierarchical nature of reality. Just as nature was discussed as being symbolic of higher aspects of reality, so then are paranormal phenomena illustrative symbolically. St. Dominic's levitation symbolizes his reaching up to heaven.

Time itself can be regarded as both paranormal and supernatural. It is paranormal in that its nature seems beyond comprehension in any orthodox scientific world view, although science, as we have seen, has found methods by which to harness mathematically aspects of time and thereby enable it to bypass the conceptual problems. Physically time is still unexplained. The supernatural aspect of time arises from its symbolic nature, pointing in the direction of a multi-levelled reality. Its origins lie above the natural, material world and through its frequently strange manifestations it enables its higher origin to be seen. In some ways time seems to be a bridge linking the material and spiritual. Like light, which itself spans between the timeless and the material worlds, time seems to be both of the world yet beyond it.

Premonition and precognition are both paranormal ways of seeing through time. Predictions of the future can also arise from prophecy, which may be either paranormal or supernatural. Just as precognition could occur in a dream, so too are there prophetic dreams. A classic example was the Pharaoh's dream in which he saw seven fat cows followed by seven thin cows, and which was successfully interpreted by Joseph. Another Joseph in the Bible dreamed that he must take the infant Jesus away to Egypt to escape Herod's treachery. Both these cases are distinctly different from each other and from the sort of precognitive dreams like those of Dunne. In the Pharaoh's dream the future prognostication was contained in its symbolism which needed interpretation to unravel it. This masking of future events within a symbolism should really be classified not as precognition but as divination, the foretelling of the future (or past or present) by interpretation of a symbolic system. The difference between the Pharaoh's dream, which essentially provided a symbolism for divination, and Dunne's dream of the volcanic island lies in the explicit vision of the precognitive and in the symbolic vision of the divinatory. Both may be paranormal, but Joseph's dream, wherein he was warned to escape, is presented in the Bible as definitely supernatural. It was a case of divine intervention over nature. Strictly speaking this latter example should be described as a vision and not a prophecy, but the division between these two categories can be indistinct.

The notion of prophecy has, of course, an Old Testament ring about it and even a comparatively modern prophet, like the sixteenth century French seer Nostradamus, seems to be part of an ancient tradition. In order to refute a prophecy by Nostradamus, his host, Seigneur de Florinville, asked the seer the fate of two piglets in his yard. Nostradamus said that a wolf would eat the white piglet and they would all eat the black one. Florinville therefore ordered his cook to kill and cook the white piglet, which he did. However, a pet wolf cub in the household began to feed on the carcass so the cook had to kill the black piglet after all to serve for dinner. Florinville announced, over the meal, that the prophecy was disproved, but, when Nostradamus contradicted him the cook was called and revealed the story in full.

In a causal framework it is hard to see prophecy as anything but a look forward along the straight line of time, but it could be a glimpse of the quality of a future time. That glimpse, which may even be a sideways glance at a part of the future that is contained in the present, could contain a feel about it in which events of a particular nature seem inevitably to become unfolded. Alternatively, the prophet may really *see* the future event as it occurs. If so, he will have jumped across time, not in the sense of a time slip, but in a similar way to the everyday experience of *seeing* a past event. It could be memory in reverse.

A supernatural prophecy that occurred in this century was the series of visions witnessed by three young Portuguese children at Fatima. In 1917 a ten year old peasant girl, Lucia, and her two cousins Jacinta and Francisco Marto, were visited by the Virgin Mary in the Cova da Iria, on the hillside above the hamlet of Aljustrel near Fatima. The three children witnessed the vision six times that year, but the Virgin Mary spoke to Lucia alone. Among other things She gave three prophecies to Lucia. The first was a vision of hell, the second was a warning of the Second World War, and the third secret of Fatima remains undisclosed. The significance of the prophecies lies in the choice they presented. War is not inevitable and can be prevented. People have the freedom to choose their destiny, that is revealed, but also the vision confirms that man can intervene, as can saints and angels, in the course of his salvation. In that case the prophet's ability to see a potential future is a very strange aspect of temporal experience. To sense an event in the future implies sensing its causes (so causality is not impaired) and the choices that lead to the future can be sensed as well as their outcome. That the choice is there enables a seen future to be averted, so the precognitive and the prophetic may actually be seeing through time to possibilities.

Up to here this enquiry has examined premonition, precognition and prophecy as ways in which time can be penetrated, seen through into the future, or at least to future possibilities. I have also distinguished between paranormal and supernatural aspects of these temporal qualities. Of course, if divine intervention can interrupt time the nature of such a penetration to the future is quite different from a natural, even though paranormal, glimpse of the future, provided by the interaction of time's own qualities and man's perception of them. I now want to turn to the fourth way the future can be foreseen and that is through divination.

10

INTERPRETING TIME

Divination means 'discovering the unknown or the future by supernatural means'. The supernatural world is a higher world, nearer the divine than the world of mankind, so obtaining divine knowledge by some process or other is seen as an act of divination. Divination is paranormal in the sense that it takes place in the natural world but runs alongside a view of normal reality, and it is supernatural in that it is concerned with knowledge from above.

Divination operates in an acausal, non-rational, symbolic world view and can only be sensibly judged within such a framework. In a view of the world where all things, indeed all events, have a symbolic significance then all things can be the subject for a divinatory system. However, the more powerful systems are based on sets of symbols which are acausally arranged with a self-consistency about them. To discover the unknown the symbols are arrayed and their arrangement at that time contains the knowledge sought. To discover knowledge of the future, to see through time, requires either a temporal sequence to the reading of the symbols or their symbolic movement in time.

Consider the complex set of symbols contained in the Tarot pack, from which modern playing cards are descended. They can be acausally arranged by being shuffled with an incomprehensible number of possible arrangements produced. A sample of the cards can then be spread out (nine is a common number used in a spread). The position of a card in the spread indicates its special significance, so, for example, the central card might represent the person consulting the Tarot, the cards to the left might represent those factors in the past that led to this moment in time and those to the right might indicate future trends. In this way the spread operates in time and the interpretation of the symbols enables time to be penetrated. Divination is an interpretive art as is prophecy. Like the chess board and its pieces, the Tarot cards form a self-consistent, all embracing set of symbols, containing in it miniature versions of all manner of human experience. The sub-set of those cards used in an act of divination then reflects the essence of that moment for that person; it contains in that moment of 'now' not just the present of here and now but also the past and the future.

The way divination works is by finding elements of the total in the local. It is as if everything in the world was like an individual cell in the human body. Locked into its form, in a language that has to be discovered, is the knowledge of every other thing; just as in each cell, with its own particular form and function, lies the genetic code that contains, locked in its own language, the knowledge of the whole organism. The aim of divination is to read the signs, unravel the symbolic language and hence obtain knowledge of the whole. As the whole contains not just spatial knowledge but also temporal knowledge, divination can lead the seer to unlock the future, to see forward in time.

Just as all things possess a symbolic meaning so too do all events, the changing relationship of things with time, have a temporal meaning. The ever shifting flux of events somehow defines each moment. In virtually all divinatory systems the quality of time is read and the current situation understood. Each moment of time contains its origins and its destiny. Hence the past and the future can be derived from an understanding of now. In most systems of divination the particularity of now is sought by some form of 'random' shuffling. The Tarot cards are shuffled, the I Ching is consulted by casting down Yarrow stalks, and they acquire the symbolic qualities of the moment. The random process, the shuffling, enables this quality of time to operate acausally. It has already been seen that randomness is probably illusory and that supposedly random processes almost certainly contain the seeds of purpose and that is what is manifest in systems of divination. If to everything there is a purpose and a time, then the diviner is finding a way to understand that purpose at that time.

Randomness and acausality lie deeply embedded at the roots of divination as well as at the basis of the quality of time, for every moment is described or clothed by the apparently random inter-relationship of all the acausally connected things around. The causally related parts of the world help provide continuity from moment to moment but the chaotic provides the feel of the time. Because each time is distinct there is always a right time for things to be done. That is, things that occur do so in harmony with the quality of the time, providing they are not forced; hence the random shuffling in divinatory processes, where the cards, sticks or whatever are allowed to pick up this temporal quality. When directed or forced to do so the system usually rebels and this is why laboratory tests of divinatory acts seldom produce a meaningful result. In the same way, if the divination does not suit someone, then trying again for a 'better' result will also be to force the situation. For example, the I Ching contains one hexagram that often turns up if someone should either doubt or try to test the oracle of the yarrow stalks. It has happened to me and to several people I know, when, in consulting the I Ching and obtaining a hexagram whose meaning is either unclear or not liked, I have tried for another hexagram. What seems almost invariably to result is hexagram four, named Youthful Folly, Mêng. Its judgement reads:

> YOUTHFUL FOLLY has success.
> It is not I who seek the young fool;
> The young fool seeks me.
> At the first oracle I inform him.
> If he asks two or three times, it is importunity.
> If he importunes, I give him no information.
> Perseverance furthers.

Divination is quite different from precognition. With the precognitive the future is seen, with divination an answer to a question is being sought. That question may be about the future, but it can also be concerned with the present or with the past. The question is really being asked of the quality of the moment in which it is posed. Jung described it as a synchronicity, the coming together in time of question, symbols and answer. Acausal nature operates in such a way that the qualitative remains distinct from the quantitative. The feel of the time contains more than its local detail, it contains a synchronism with the whole of time. A moment of time is then like a virtual particle, appearing out of the vacuum of time and space yet interconnected with the whole universe.

A simple divinatory system that most directly illustrates the nature of all such systems and their dependence on questioning is dowsing. Most dowsing is connected with place not time, yet dowsing can work into time past or future. Similarly, dowsing does not have to operate at a real location but can be done from a map. In map dowsing, what is happening is that the symbolic representation of a natural location is as effective a symbol as the real location itself. The symbolism of the thumb print, to the Tibetan monk, is as powerful as a real map in helping him find his way, when lost, across the snow covered mountains. The difference in dowsing from nature or from a map is only one of degree. The map is a secondary symbolic system, it is a shorthand version of part of nature, which itself is a symbolic version of a higher reality. Such a view does not deny the 'real' nature of the physical world, but expresses its relationship to a greater reality. Dowsing, therefore, becomes not a sensitivity to the local environment but a mechanism for asking divinatory questions of a very simple nature (basically, questions that can be answered yes or no). Dowsing is not referred to as water *divining* by chance, its name is an accurate description of what it is.

An anecdote, told by Tom Graves, concerning dowsing from a map not only illustrates the rare power with which such a simple divinatory system can operate, but also shows how dowsing can see through time by questioning the moment of now. A dowser had a friend who was always on the move, and he wished to contact him. He phoned the two numbers he had but from both places the friend had departed. The dowser took a pendulum and an atlas of London and started on the index map. One page stood out clearly so he turned to that page and one street then emerged. Could the friend be at a hotel, the dowser wondered. He took out a trade directory and found a list of hotels for that street. One hotel gave a distinct reaction so he phoned the hotel and asked for his friend. He was told there was no one there of that name. The dowser was puzzled and left a message with the hotel receptionist 'just in case'. Two hours later the friend telephoned. He was at the hotel and decidedly curious as to how he had been located, particularly as he had only decided to go to that hotel one hour before!

This example shows how complex and interwoven time can be. The present moment contains past and future and past within the future. The dowser's attunement with time enabled him to trust his hunch, to believe his answer even though that future had not yet become a fixed reality in the present. The future can be different from the precognition of it and the potential future foreseen in divination may not emerge in quite the same way. When people travel on a journey they cannot alter their already trodden path, but their destination can be changed even if it is constricted by the routes that are available to certain possibilities. So it is on a temporal journey.

The crux of the act of divination is the posing of a question. The water diviner is constantly asking the question: 'Is there water there ... or here?'. If the question is a genuine one an answer will be given by whatever symbolic system is being consulted. If the question is not in harmony with the quality of the time then the answer may not 'make sense' or be very obscure. In some divinatory systems, such as in horary astrology, the oracle itself may not be open to consultation, in that certain configurations of symbols mean 'no answer can be given to this question; the time is wrong'.

Asking a question is an active process, yet the underlying assumption in divination is that the answer is somehow passively there, waiting to be released. This is not altogether an accurate picture because the answer to a question asked at a particular time will only be available at that time. It is as if it were there waiting. The only time such answers shout out actively is through an omen, a sign whose divinatory message needs interpretation, but does not require a question. Although the question is not explicit it could be argued that omens appear when people are in need of posing a question.

The paradox of time displayed by divination centres round the containment of time past and time future in the present, the moment of now. Another paradoxical face of time lies in its earthbound relentlessness, the passing seconds, hours and days, contrasted with its ethereal timelessness, its intangibility. The divinatory system that is perhaps the most complex and intriguing of all also displays a curious temporal paradox. The system is astrology and its paradoxical aspect relates to the precision with which it operates in time, placing events to the nearest minute or even second, while also using time in a purely symbolic sense.

Astrology as a divinatory system enables knowledge of the unknown and of the future to be acquired through the interpretation of the symbolism of the planets, their positions in the Zodiac and in the astrological houses, as well as in their detailed interrelationships, which themselves change with time. Unlike other systems, astrology does not proceed via a random shuffling process but focuses the quality of time into a horoscope, which is cast for what is effectively a 'random' time and place, that is the moment of birth. The map of the heavens then produced for that moment and for that place (every such chart being unique) represents the quality of time that synchronistically is mirrored from the celestial pattern above to the terrestial event below.

To set up a horoscope requires knowledge of the time and place for which the horoscope is being erected. If the chart is for someone's birth, a natal chart, then that information will relate to that person, but horoscopes can be constructed for any event in space and time. The date yields information about the position of the planets in the heavens, which is usually discovered from astronomical tables called ephemerides. The time of day is required both to ensure the accuracy of planetary positions (the moon moves about one degree in the sky every two hours) and to set up the system of astrological houses. At the exact moment of the event in question a particular part of the zodiac will be rising on the eastern horizon. This is taken as a reference point for the horoscope and is referred to as the ascendant or rising sign. When the ascendant has been determined the chart is then divided into twelve houses, starting with the ascendant as the beginning of the first house. The houses do not, in general, coincide with the signs of the zodiac, but mirror the qualities assigned to the signs. Similarly, there will be a particular sign overhead which, with the ascendant, marks out the east/west and north/south axes of the map. The sun, moon and planets, which all appear to move around the earth in their cyclic paths, are then placed onto this chart in their correct astronomical positions. As they all move in essentially the same plane, the plane of the ecliptic, they appear in the twelve signs of the zodiac, which have the same names as the constellations found along the ecliptic. The place of the event is also required for determining the ascendant because different parts of the zodiac are visible from different longitudes and because the heavens move, with respect to any observer, at different rates at different latitudes. For example, the sun is visible for twelve hours a day at the equator, while in northern latitudes it may be visible for only eight hours in the winter but for sixteen hours in summer months.

In this way every place and every time has a unique mapping onto a horoscope which then forms a symbolic representation of the event which it mirrors. It is then accessible to interpretation. This should be an act of divination by the astrologer, who has come together in time with the map. This encounter itself needs an auspicious time, a harmony, to become a truly divinatory act. Most often astrologers are found interpreting maps in a casual fashion, in that it matters not what the map is, what the time is, and often without that attitude of mind that is necessary for divination, a necessity of purpose and a question to be answered. Astrology or divination is an act of religious import. Otherwise all that can be done is to play with symbolic meanings very much like a game, and this, I am afraid, is what many astrologers seem to do.

Astrology is an acausal system, even though it contains the continuity of the causally connected motions of the planets. To search for its origins and power by looking at those causal parts, is of course a mistake and an ignorant mistake, and one that would be less easily made with other, more obviously random divinatory methods. Astrology works because each planet, each angular interrelationship or aspect between the planets, each zodiacal sign, each house and even each degree of the zodiac, all have special qualities or essences, that, once known, can be interpreted. The divinatory act is to bring together all these symbolic configurations into a whole picture of the quality of the event for which the map has been cast. Within the map will be the answer to the question sought. The most complex form of astrology is, therefore, natal astrology, interpreting a birth chart, for that chart contains the answer to the very complex question 'what sort of person is this?'

Whereas many critics of astrology attack it on the grounds that there can be no sensible causal connection between a person on earth and the actual physical planets, the truth of the situation lies in the acausal correspondence between an apparently random moment in time, the time of any event, and the exact but relative positions of the planets at that time. There is no causal connection. The power of astrology as a divinatory system lies in its acausality, just as in Tarot the shuffling of the cards is the vital acausal link. The random moment of birth, even the random pattern of tea leaves in a cup, has qualities and an interconnectedness with the rest of nature, because nothing is really random in the sense of being meaningless.

It is ironic that this view is in contrast to the ever more widely accepted view that emerges from the purely quantitative instructional sciences. Nobel prize winner Steven Weinberg has written: 'The more the universe seems comprehensible the more it also seems pointless.' But such a view does not emerge if qualities are not reduced to quantities, if the acausal is not lost sight of.

Before considering in what time astrology may actually operate, I would like to present an example of precise astrological timing and an example of the symbolic use of time. The first of these involves a royal occasion and is yet another extraordinary anecdote involving clocks and watches. The event in question was subject to precise timing by two independent astrologers. The occasion was the wedding of Princess Anne to Mark Phillips. The moment chosen by both astrologers, independently of each other, to signify the marriage was the phrase 'I pronounce you man and wife, in the name of the *Father* ...'. The word 'Father' was chosen as the moment at which the marriage becomes irreversible, becomes 'made in Heaven'. As the Archbishop said the word 'Father' the sun touched the mid-heaven exactly to the second. Other resonances also sounded in the horoscope for that moment, but let me quote what also occurred to one of the astrologers, Geoffrey Cornelius.

Now an extraordinary twist: my old watch was reading the wrong *minutes* as the second hand went to 45 seconds, because minute and second hand were independent. I initially therefore got the right *second* exactly one minute out and cast the horoscope accordingly. (I had already, and I did immediately after the timing, check against TIM to adjust my incorrect seconds reading). On casting the map it struck me that the sun was exactly one minute in time from culmination and *I immediately guessed that my reading of the watch was incorrect*. I checked carefully against TIM (the Greenwich time signal), watching for the ambiguous reading of the minute hand, and at what second this occurred, and affirmed what the map had already suggested to me. Here is an extraordinary example of correcting a time-piece from the astrological interpretation of events.

The correct timing was confirmed independently by the other astrologer. This example shows the power of the astrological effect to concentrate correspondences of events and their symbolic representation through planetary positions in time. The uniqueness and complete individuality of the event is also clearly illustrated, which renders it accessible to descriptive but not to instructional science. A recipe cannot be written for a royal wedding with astrological significance, but when a unique event occurs like this the quality of the moment comes together symbolically and physically.

A horoscope is not a static, passive thing but dynamic. The pattern of the planets in their placings for the moment in time for which the map was cast marks out that one moment, but the planets also move, not just in the heavens but on the chart. There are a number of ways in which a horoscope changes with time; these are collectively called progressions. For example, the system of secondary progressions takes one day in a life to be symbolic of one year. To progress a chart by this method means seeing where planets had moved to, say, twenty days from birth; their positions then symbolize how that person will have developed after twenty years. In addition an astrologer will take into account, in a detailed analysis, the positions of the planets actually in the sky after twenty years and their relationship to the natal planets on the chart. These correspondences between now and the moment for which the chart was constructed are called transits.

The notion of moving the symbols of the planets on a horoscope is itself completely symbolic. It does not relate either to clock time or to the real positions of planets and yet the method is uncannily powerful. A well studied example comes from Mussolini's life. On 30 October 1922 Mussolini 'marched on Rome' in a bid for power resulting, on that evening, of the King conferring on him full political leadership. His progressed planets for that time in his life struck up a harmony such that the relationship between the Moon and Jupiter coincided, to the day, in a symbolic rapport that signified his relations with the public being benefited by his emergence into authority. At the very moment that Mussolini saw the King the sun also transited the point on his birth chart known as the North node which signifies future destiny and influx of power.

It is difficult to convey the powerful meaning inherent in astrological symbolism and yet like other languages it is both vital and communicable. Another example, and one I was personally involved in, illustrates the universality of astrological symbolism well, but first it needs a brief introduction to that branch of the subject called horary astrology. Horary means 'concerning the time' and in the practice of horary astrology the person seeking an answer to a question poses the question to the astrologer. A map is erected for the moment the question is asked and the answer to the question will be contained in the map, providing the map can be read. Horary astrology is relatively simple and is bound by a series of straightforward rules, which when followed lead to a judgement. The first rule concerns the map's readability with certain configurations of the planets indicating the chart's invalidity. The example I shall give of a horary reading comes from an event.

I have already mentioned the omen of seeing a flamingo in an English country field. The incident occurred to me when lecturing at a residential course, whose theme was 'Appearances', which involved scientists, astrologers and Buddhists.

A chart for the moment the course began contained only one interesting feature, namely the exact conjunction of the Sun and Mercury which would take place on Saturday afternoon at about 3.10. At that time, but unaware of its significance, we saw the flamingo. Also at that time another group of people at the course discovered a drowned sheep, stuck between some rocks on the sea shore. They too had noted the time of this omen. The chart for these 'appearances' included, of course, the conjunction of the Sun and Mercury, but it also turned out that the conjunction was exactly on the point at which the ninth house began, and the conjunction made a significant angular aspect to the Moon at that moment. The horary reading of that chart, in answer to the question 'what will be the nature of the appearance at that time?' can be given in this way. The Sun conjunction was in Aries, the ram. Aries is ruled by the planet Mars, which was found in the eighth house, which signifies death and rebirth, and it was in the sign of Pisces, the fish. Here was written large, a ram, dead in water – the drowned sheep. As for the flamingo, as someone pointed out, a flamingo is the nearest living thing we know of to the phoenix, the symbol of rebirth.

The synchronicity of this event with its astrological portrayal, also linked the collective idea of 'appearance' to the group. The map of the event was shown blind to a major astrological conference whose participants were asked to suggest the nature of the appearance, just from the symbolism of the map. The same judgement was given and in such a way that when the 'drowned sheep' theme of the chart was revealed the audience was moved by the stark simplicity of the astrological effect, which so often seems confusingly complex. That moment of time, reflected in the planetary positions, presented its form to us through two omens. The quality of time did not need to be questioned but was demonstrated with clarity.

The relationship between divination and time is far less clear than was the case for precognition. It seems that every moment contains the essence of all other moments within it, just as one particle contains within it the essence of all other particles, or that the moment of here and now in space and time also contains the past and future connected to it by light. However, as the last example demonstrated, there is also a connection between the moment and the individual experiencing it. The dead sheep and the flamingo were at the places where we found them for many moments but their significance for us was held in the exact moment that we saw them.

The nature of time in which astrology operates is also confused by its precise operation in clock time. The time, the objective time maybe, of an event, or when a question is posed, contains that event's significance. The symbolic movement of the qualities of that time, contained in the planetary positions, also reveals how the judgement changes in time. It is a symbolic movement of time that reveals past and future from a moment of present time encapsulated at a precise instant of clock time. Symbolic time links to real time as its source and as its goal. However, a final example denies astrology even this basis in objective time.

In 1975 the Humanist magazine in the U.S. published a three pronged attack on astrology. As well as two articles, written by scientists, arguing against astrology, the magazine published a statement condemning astrology as having 'no scientific foundation' and is 'based on magic and superstition'. The statement was signed by 186 leading scientists, mostly astronomers, including eighteen Nobel prize winners. I am not going to discuss that attack here except to comment that, like Feyerabend, I am certain that most of the scientists signing the document never investigated astrology, but based their opinion purely on their own superstitions about it.

One of the articles accompanying this statement was written by the astronomer Bart Bok. In his article Bok included a horoscope, drawn up for a date in 1907 and based on New York. The map was taken from an article he wrote with Margaret Mayall in 1941 in which they both attacked astrology. It is presented as a typical birth chart, although why this map was produced is not known. It is not

Figure 10.1 *The horoscope presented in the* Humanist *attack on astrology which signifies 'An Attack on Astrology'.*
Drawn up by Bart Bok and Margaret Mayall for 23 November 1907, 08:59:52 Universal Time, New York City 40N43; 73W58. Topocentric cusps.

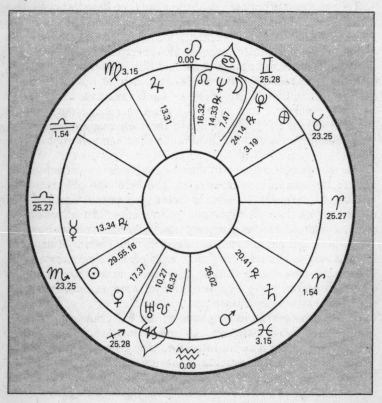

the birth map of either Bok of Mayall and may be entirely arbitrary. It is reproduced here as Figure 10.1.

What is extraordinary about this map is that it can be read, not as a natal chart at all, but as a symbolic representation of an attack on astrology. If the map is treated as providing an answer to a question about the state of astrology then its response is clear. Not only are the symbols for astrology and astronomy on the map in conflict but the nature of the attack itself is indicated. There is powerful meaning which speaks out to any traditional horary astrologer in the pattern of this chart. The reading continues to tell us that the attack will not be successful. Its perpetrators will be 'sceptical in religious matters, their reason and faith are in conflict and their judgement is poor'. The symbols also show that the astrologers against whom the attack is being made are too dumb to defend themselves and are 'not likely to win credit for themselves or their art'.

The message of this map was first grasped by Geoffrey Cornelius, who said that it describes the 'obscuration of the core issue of astrology'; it is to be read as not just 'an attack on astrology' but for the details it gives of the nature of the attack and the state of the astrology being attacked. I have examined both the chart itself and Cornelius's analysis of it and agree that its language communicates this message plainly. Furthermore, the progressions and transits, working from a map cast for 1907, strike important chords for the 1941 usage and even more dramatic resonances for the September 1975 Humanist attack. The map operates in symbolic time quite clearly, placing the attack into its correct position in clock time.

The implications of this chart are remarkable. It was chosen at random by two people who thought it showed how silly astrology was and yet it was in fact a description of the attack for which it was used. This 'random' chart not only progressed to show the serious renewal of that attack in September 1975, but it also progressed to 1941 when it was first used in attacking astrology, thirty-four years after it was effectively cast. For astrologers this horoscope is important because it illustrates that not only does the moment of birth contain meaning but any moment can, even a random moment chosen at will. The astrologers' art does not only occur in 'objective time' but occurs in a temporal symbolism, a synchronicity, where events come together with meaning, with divine knowledge. The map points clearly to the divinatory nature of astrology.

For this enquiry into time the map is also important. It seems to bring together the ideas of randomness, coincidence and divination into one place. The moment of time which encapsulates not just its own quality, but also the quality of a separate earlier time, which in turn encapsulates both the later time and yet another time in which all three times reach a fulfilment of meaning, is past comprehension. What this astrological map demonstrates more clearly than any other example I know is that *all* time is symbolic, even the sort of time called 'objective' or 'clock time'. The paradox of astrology is a paradox resolved when all time is seen to be symbolic. Time seems to touch mankind materially, physically in everyone's lives as clock time, but in many other ways, often unseen. Time seems to touch man acausally from somewhere other than the material world of clocks.

11

THE PARADOX OF
TIME

This enquiry into Time began by asserting time's existence. The journey pursued in these pages has been an attempt to track down time's elusiveness. As was suspected at the beginning time has still escaped but it is now, I hope, more clear what sorts of questions can be asked of it. This is not the end of the quest, but merely a resting place. One thing now known is that time has made its mark on us, as it does on every journey we embark on. Like all other journeys this one has not really ended, because travellers always get up and move on again, at least in this mortal life. But that is merely because of time's encompassing yoke.

Having reached this resting place, can time be explained? It cannot, but rather it is clear that time is part of human experience, of very existence. The problem is more one of explanation in preference to experience. Our curse, maybe the curse of time, is to always try to explain. What remains are answers to questions, phrased in terms bound by the nature of those questions. Those questions are like prisons, the answers give descriptions that are prisoners. They serve their time until the question is changed and another description, another prisoner, is created. If we see only our questions and our descriptive answers, are we not also prisoners of time? Time is our captor. We can only glimpse out of our prison to see a limited view of events unfolding: which is time.

But if bodies are constricted in time, thoughts and spirit may be able to at least envision out beyond it. Man's imagination may take him, via the world of experience, to question the very nature of his captor. In prison he can escape in thought, rise above the condition and in many different ways that is what we have been doing on the journey through these pages. We have broken some of the chains of time. What has been released has been a complex thing whose multifaceted displays have often appeared paradoxical, but usually because we have asked the wrong questions, collided with the prison walls again.

The world of time has appeared as change. Things change, events occur, time moves on, the river flows. Our clocks operate at different rates, each one of us has our own time, physically and psychologically. We experience time in different ways and at different rates. When we move so too does our proper time change in relation to others.

We have seen that man can somehow slip in time, into different times. We can remember back in time and seem to be able to see outward, forward in time, or else time itself reaches back to us. Dramatic events may even splash in a temporal matrix that extends beyond now into past and future. We may ourselves extend in time, forward and backward. Time has paradoxically presented us with an objective face of physicality while also hinting at its lack of dimensionality. Time as change on a linear scale extends forever, seems so real, tangible, and yet we have seen the face of time wherein all times are now; past, present and future all coinciding.

The riddle of time and its many manifestations in thought and experience seems somehow to tease us, embedded as we are in our own particular place in time itself. We are so imbued with the quality of our age that time's many facets seem almost too unbelievable to our rigid minds. Yet time is magical. Its interpenetration of the fabric of the universe, its eternal presence in now, its paradoxical nature is nothing less than magic. Not the magic of the conjurer, performing tricks, but magic in the sense of the wisdom of the spirit. Its nature is occult, hidden from us by its own disguises, its physical display. It eludes the scientist, confounds the philosopher and, like Puck himself, is found here, then there, never where we are looking but never far away.

We have found time to be symbolic, but also at the roots of causality. We have found it at the basis of the quantum description of the micro-universe and at the heart of cosmology. We have located time in the clocks that are used and yet time has been described as paranormal. Its magical capacity to transform itself into something else lies in its own essence.

T.S. Eliot described this perhaps best of all in 'Burnt Norton'.

> Time present and time past
> Are both perhaps present in time future,
> And time future contained in time past.
> If all time is eternally present
> All time is unredeemable.
> What might have been is an abstraction
> Remaining a perpetual possibility

> Only in a world of speculation.
> What might have been and what has been
> Point to one end, which is always present.

Eliot's verse contains so many threads that have been found in this journey that to identify them all would be too long a task. The eternal moment of now, containing all past time and all future time rings out not just in the paranormal and divinatory manifestations of time that have been examined but also in the scientific analysis. The interpenetration of all things with everything else, revealed in the world of virtual particles, revealed by the role of time in cosmological theory and most of all displayed by light, ties together the scientific view with a mystical, a non-rational view so far excluded by science.

The interpenetration of space and time is comprehended in theory. We see, however, not that interpenetration, that timeless unity in the universe, inside ourselves. What we see are physical effects, manifestations in real time and real space of that interpenetration. It displays its spaceless, timeless nature through time and space. The only glimpses of it otherwise are through the mediating influence of light, and possibly in the paranormal, supernatural events that come into our lives on rare occasions.

The interpenetrative moment of now, in which all time is eternally present, connects everything. And so does light. Light is itself timeless, as we have seen, but not eternal. Light is in this world of manifestation and yet it is not in the world of time. Time derives from light. Without light there would be no time, and yet light itself is timeless. Light touches us in time, connects us with all other times, and in its touch both ties us to time and frees us from it.

Krishnamurti said that time is the distance thought travels; the scientists measure distance by the time light travels. To St. Thomas' angels time has duration but no sequence, just as light has sequence but no duration. Light is a manifestation of spirit, but one that makes connection with the mortal world in an overt way. It has been argued that angels themselves direct time from their timeless realm. Light governs time on earth, clocks are set by it, yet light, like the angelic realm, is timeless. Does time 'point to one end, which is always present'? Is that eternal present the timeless realm of light and of angelic hosts? In God's creation and in modern man's interpretation of it in cosmology, light comes first, then matter.

Light has more primacy, is higher up the ladder, so to speak. One end of light is timeless, our end of it imbues the world in time. To us its speed is finite, to itself it is infinite. Puck could encircle the world in forty minutes but light, in its realm, can encircle the universe in an instant. Light symbolizes spirit and that symbolism can be seen even in a scientific context. Our discussions about light and time lead us to see one end toward which it points, to its symbolic meaning.

In the realm of light causality disappears. Cause and effect are linked instantly in a timeless dimension, yet light's appearance in the physical world is at the root of causal connection. Light again is the bridge that links us with a meaning. The significance of causality itself is in the continuity it provides and with the insight that its roots, in light, are causeless yet caused. Time also, causeless but caused by light, lies at the roots of causality. Angelic time, which lacks sequence, lacks causality, is an acausal time. The corporeal world too has its acausal side. Again we find duality, causal and acausal sharing the same space and time, both faces of the complexity of our reality.

The world of acausal phenomena, the world around us of unconnected and yet interlinked events, has been bypassed in the scientific description of that aspect of the world that displays lawful generalities, that is the causal side. Instructional science cannot handle acausalities, just as it cannot handle individual events. But we have also seen a description of reality in which all things are individual and unique. The direction of instructional science is itself paradoxical, in that its increasingly detailed search for ordered principles in phenomena leads eventually to each thing studied displaying its own individuality. The transition from mechanistic laws to statistical ones in modern scientific analysis is an indication of this dilemma, for the statistical descriptions are nothing else but generalized patternings of individual events. In such a method of description, however, the patterned, the apparently causal side of things, alone emerges, and the qualities of the individual things themselves is lost. The acausal side of nature is disregarded, and that is our loss.

Yet time also illuminates the acausal side of the world. Temporality necessitates both causal and acausal, the patterned and the random, but in all its aspects is found meaning. Chance itself is not meaningless. The basis of divination lies in chance and acausality. Present day methodology overlooks the significance of the acausal and gives importance to the less important side of nature. A literal view of the world has meant that signposts are seen as signposts and not as pointers toward something else. We are all dogs biting the fingers that are showing us where we should be going (ending up as statistics on dog related injuries!). Instructional, statistical science has got its priorities mixed up. It has missed the point and subsequently found the world increasingly pointless. That alone is surely a clear enough indication of how science has lost touch with human experience. We still prefer to base our judgements on the theoretical rather than the experiential, leading us further into blind alleys, back into our prison.

Links do exist between the scientific and the apparently fantastical. The minute gaps in time, maybe the particulate composition of quantum time, which was found in the quantum description of the sub-atomic world enables matter to become manifest in virtual reality. Such a notion as virtuality at the roots of the material world, echoed so well by Eliot as 'perpetual possibility ... in a world of speculation', is indeed fantastical. More so maybe than the notion that for fairies three hundred human years appear as three days. At least conciliation is found in this latter idea in that it sounds very much like Einstein's riding a light beam at the initiation of his concept of relativity.

I am sure it is mistaken to take the apparent equation between virtual particles and ghosts, or between the sub-quantum vacuum and the ethereal world of spirit too seriously, for such equalities only appear out of an instructional world view, but there is some appeal in such equations. If those analogies are pushed too far, however, we will inevitably end up against a brick wall, cornered, so to speak by snarling dogs who want to bite our pointing fingers. Much safer then to choose a different path, to describe the world in a different way, for the world exists for us only in the form we clothe it in. Our descriptions or explanations define our world. Our technology manifests our explanations. Its soullessness only highlights the pointless nature of an instructional universe.

We have seen, in this journey, that reality is defined and redefined in our discovery of it and in the way in which we circumscribe it with explanations. Reality was redefined with the introduction of the clock. Time was redefined mechanically. But the clock itself is symbolic, not in a literal sense of indicating the time of day via its pointers and numerals, but as an entity symbolizing time, the time man created and created, paradoxically, in the form of the clock. Clocks have weaved their way into our journey. We have encountered surrealist clocks, Dunne's watch reading 4.30 in the afternoon or night; Alec Guiness misread two alarm clocks after sleeping through both their rousing calls and Geoffrey Cornelius corrected his watch by a moment of significance in a wedding, where the clock of heaven was more accurate than his mechanical time piece.

These incidents with clocks (these pieces of invented time) all suggested things were not right with the way time was defined. They again were pointers showing us the error of defining time objectively. Clock time is invented time, but man has been too gullible, he has ended up believing that his invention has an objective existence separate from the way he designed it to be. No wonder there is confusion about whether time is merely a matter of consciousness, of whether time exists only in the mind. Objective time, clock time, exists because the mind invented clocks. That invention gave us a definition of an apparently objective time that we believe in too much. If we did away with clocks, objective time would also disappear. Time would then be redefined in ways that suited our existence in a clockless world.

Objective time has gone. It has gone in relativity, gone from the quantum world, gone in cosmology, where scales and redefinitions of time are almost arbitrary. Objective time has vanished in the paranormal. Only in the 'normal' world, which has been impoverished by our definitions and explanations which define poorly and explain little, does objective time still hold sway. One thing seen in this journey is that the time we think we inhabit is neither simple, linear nor objective. We are immersed in pools of time and timelessness, in a sea of causality and acausal connections. We are flooded by light that brings us time and banishes it.

Time has shown the limitations of description, the extent to which they have validity. Time has shown the limitations of science, of language, the limits of the world. Yet time has shown us beyond the world, for, like all creations, time acts as a pointer to a greater reality than the one we have drawn around ourselves. Time is a pointer to the symbolic nature of reality. Time is symbolic itself and makes sense of itself and all phenomena in a symbolic reality. Beyond time is light and everything that light itself symbolizes. Light floods us with illumination so that we can see. Seeing is itself a symbol, a pointer, for we see in order to see the origin of light and the origin of time.

This journey has taken us from the material world, the world that belongs to human experience, and has brought us as far as a signpost pointing to the more complex nature of reality. The signs are pointing to the worlds above ours (in a symbolic sense), the worlds of spirit. In bringing us this far time has not gone away. We have not pinned it down, it remains elusive. Rather it has led us along our path, a path strewn with paradoxes, with fantastical phenomena and strange descriptions, but a path bound to our mortal state. Beyond mortality time may escape us altogether or we may find it by losing it. In its discovery, in the revelation of its secret, it will remain paradoxical. Our guiding torch, time and light together, has brought us far. We are resting here but we will inevitably continue. Time does not stand still, at least for long.

The complexity of time enfolds us. It leads into experience and away from explanation. We still try to explain – which also is bound up in our mortality, in our timeboundness, but St. Augustine knew that time's experience was more real than its explanation. Where time leads us we must follow, where it points to we may reach. Some time.

BIBLIOGRAPHY

GENERAL

L.R.B.Elton and H.Messel: *Time and Man*. Pergamon, Oxford, 1978.

R.M.Gale (Ed): *The Philosophy of Time*. Harvester, Sussex, 1968.

J.Grant and C.Wilson (Eds): *The Book of Time*. Westbridge, Newton Abbott, 1980.

W.H.Newton-Smith: *The Structure of Time*. Oxford University Press, Oxford, 1980.

J.B.Priestley: *Man and Time*. Aldus/Allen, London, 1964.

S.Toulmin and J.Goodfield: *The Discovery of Time*. Hutchinson, London, 1969.

G.J.Whitrow: *What is Time?* Thames & Hudson, London, 1972.

P.J.Zwart: *About Time*. North-Holland, Amsterdam, 1976.

CHAPTER ONE

P.C.W.Davies: *Space and Time in the Modern Universe*. Cambridge University Press, Cambridge, 1977.

T.Gold (Ed): *The Nature of Time*. Cornell University Press, Ithaca, 1967.

D.R.Hofstadter: *Gödel, Escher, Bach*. Harvester, Sussex, 1979.

M.McLuhan: *Understanding Media*. Routledge and Kegan Paul, London, 1964.

L.Mumford; *Technics and Civilization*. Routledge and Kegan Paul, London, 1934.

R.Sheldrake: *A New Science of Life*. Blond & Briggs, London, 1981.

J.Weizenbaum: *Computer Power and Human Reason*. Freeman, San Francisco, 1976.

CHAPTER TWO

F.W.Cousins: *Sundials*. Baker, London, 1969.

N.Feather: *Mass, Length and Time*. Penguin, Harmondsworth, 1961.

R.Hanbury Brown: *Man and the Stars*. Oxford University Press, Oxford, 1978.
D.Howse: *Greenwich Time*. Oxford University Press, Oxford, 1980.
A.L.Rawlings: *The Science of Clocks and Watches*. Embury Press, Wakefield, 1974.

CHAPTER THREE

H.Bondi: *Relativity and Commonsense*. Heinemann, London, 1964.
J.Coleman: *Relativity for the Layman*. Penguin, Harmondsworth, 1954.
A.Einstein: *Relativity*. Methuen, London, 1960.
A.Einstein & L.Infield: *The Evolution of Physics*. Simon & Schuster, New York, 1938.
W.J.Kaufmann III: *Relativity and Cosmology*. Harper & Row, New York, 1977.
J.Narlikar: *The Structure of the Universe*. Oxford University Press, Oxford, 1977.
J.Pasachoff: *Contemporary Astronomy*. Saunders, Philadelphia, 1977.
D.Sciama: *The Physical Foundations of General Relativity*. Heinemann, London, 1969.
A.M.Young: *The Reflexive Universe*. Delacourt, New York, 1976.

CHAPTER FOUR

J.Andrade e Silva and G.Lochak: *Quanta*. World University Library, London, 1969.
S.W.Angrist and L.G.Hepler: *Order and Chaos*. Basic, New York, 1967
F.Capra: *The Tao of Physics*. Fontana/Collins, London, 1976.
A.Eddington: *The Nature of the Physical World*. Dent, London, 1928.
R.M.Eisberg: *Fundamentals of Modern Physics*. Wiley, London, 1967.
J.F.Kirkaldy: *Geological Time*. Oliver & Boyd, Edinburgh, 1977.
I.Prigogine: *From Being to Becoming*. Freeman, San Francisco, 1980.
E.Thomas: *From Quarks to Quasars*. Athlone, London, 1977.

CHAPTER FIVE

J.D.Barrow and F.J.Tipler: 'Eternity is Unstable'. *Nature*, vol. 276, p. 453, 1978.
V.M.Canuto: 'Does Gravity Vary With Time?'. *New Scientist*, 15 March, 1979.
A.Eddington: *The Expanding Universe*. Penguin, Harmondsworth, 1940.
J.Gribbin: *Timewarps*. Dent, London, 1979.

F.Hoyle: *Astronomy and Cosmology.* Freeman, San Francisco, 1976.

F.Hoyle: *Steady State Cosmology Revisited.* Cardiff University Press, Cardiff, 1980.

J. Rosen: 'The Extended Mach Principle', *American Journal of Physics*, 1981.

H.Shapley: *Galaxies.* Harvard University Press, Cambridge, Mass., 1972.

S.Weinberg: *The First Three Minutes.* Deutsch, London, 1977.

CHAPTER SIX

D.Bohm: *Causality and Chance in Modern Physics.* Routledge and Kegan Paul, London, 1957.

D.Bohm: *Wholeness and the Implicate Order.* Routledge and Kegan Paul, London, 1980.

H.C.Dudley: *Morality of Nuclear Planning.* Kronos, Glassboro, New Jersey, 1976.

R.Duncan & M.Weston-Smith: *Encyclopaedia of Ignorance.* Pergamon, Oxford, 1977.

A.Hardy, R.Harvie and A.Koestler: *The Challenge of Chance.* Hutchinson, London, 1973.

A.T.Oram: 'An Experiment with Random Numbers', *J. Soc. Psychical Research*, vol37, p369, 1954.

G.Spenser-Brown: *Probability and Scientific Inference in Scientific Investigation.* Longman, London, 1957.

W.Weaver: *Lady Luck.* Penguin, Harmondsworth, 1977.

CHAPTER SEVEN

J.C.Cooper: *An Illustrated Encyclopaedia of Traditional Symbols.* Thames & Hudson, London, 1978.

M.Douglas: *Natural Symbols.* Penguin, Harmondsworth, 1973.

J.W.Dunne: *An Experiment With Time.* Faber, London, 1934.

C.G.Jung: *Synchronicity.* Routledge and Kegan Paul, London, 1972.

P.Kammerer: *Das Gesetz der Serie.* Verlags-Anstalt, Berlin, 1919.

A.Koestler: *The Roots of Coincidence.* Hutchinson, London, 1972.

L. van der Post: *A Mantis Carol.* Hogarth, London, 1975.

E.F.Schumacher: *A Guide for the Perplexed.* Cape, London, 1977.

CHAPTER EIGHT

R.Collin: *The Theory of Celestial Influence.* Stuart & Watkins, London, 1971.

J.Cohen and J-F.Phipps: *The Common Experience*. Rider, London, 1979.
Krishnamurti: *Penguin Krishnamurti Reader*. Penguin, Harmondsworth, 1970.
M.Manning: *The Strangers*. Allen, London, 1978.
R.E.Ornstein: *On the Experience of Time*. Penguin, Harmondsworth, 1969.
A.P.Shepherd: *The Eternity of Time*. Hodder & Stoughton, London, 1941.
C.Wilson: *Mysteries*. Hodder & Stoughton, London, 1978.

CHAPTER NINE

E.Cheetham: *The Prophecies of Nostradamus*. Spearman, London, 1973.
Fr.L.Kondor (Ed): *Fatima in Lucia's Own Words*. Postulation Centre, Fatima, 1976.
C.S.Lewis: *Miracles*. Collins, Glasgow, 1947.
M.Manning: *In the Minds of Millions*. Allen, London, 1977.
K.Pedler: *Mind Over Matter*. Thames/Methuen, London, 1981.
J.Taylor: *Science and the Supernatural*. Temple Smith, London, 1980.
C.Wilson: *The Occult*. Hodder & Stoughton, London, 1979.
C.Wilson and J.Grant: *Directory of Possibilities*. Webb & Bower, London, 1981.
J.White and S.Krippner (Eds): *Future Science*. Anchor, New York, 1977.

CHAPTER TEN

B.Bok, L.E.Jerome and P.Kurtz: 'Objections to Astrology'. *The Humanist*, Sept/Oct, 1975.
C.E.O.Carter: *The Astrological Aspects*. Fowler, Romford, 1967.
G.Cornelius: 'An Anti-Astrology Signature'. *Astrology*, vol. 52, no. 3, p. 88, 1978; vol. 53, no. 1, pp. 5–11, 1979.
P.Feyerabend: *Science in a Free Society*. NLB, London, 1978.
A.Douglas: *The Tarot*. Penguin, Harmondsworth, 1972.
T.Graves: *Dowsing*. Turnstone, London, 1976.
M.E.Jones: *Horary Astrology*. Shambala, San Francisco, 1943.
D.Rudhyar: *The Astrology of Personality*. Doubleday, Garden City, 1970.
M.J.Shallis: 'Astrology and Science'. *Astrological J.*, vol. 22, p. 42, 1980.
R.Wilhelm (trans): *I Ching*. Routledge & Kegan Paul, London, 1951.

INDEX